METEOROLOGY MANUAL

THE PRACTICAL GUIDE TO THE WEATHER

First published in April 2014.

A catalogue record for this book is available from the British Library.

ISBN 978 0 85733 272 1

Library of Congress control no. 2013952658

Published by Haynes Publishing,
Sparkford, Yeovil,
Somerset BA22 7JJ, UK.
Tel: 01963 442030 Fax: 01963 440001
Int. tel: +44 1963 442030
Int. fax: +44 1963 440001
E-mail: sales@haynes.co.uk
Website: www.haynes.co.uk

Front cover images courtesy of Davis Instruments and Wikimedia.

Haynes North America Inc.,
861 Lawrence Drive, Newbury Park,
California 91320, USA.

Printed in the USA by Odcombe Press LP,
1299 Bridgestone Parkway, La Vergne,
TN 37086.

METEOROLOGY MANUAL

THE PRACTICAL GUIDE TO THE WEATHER

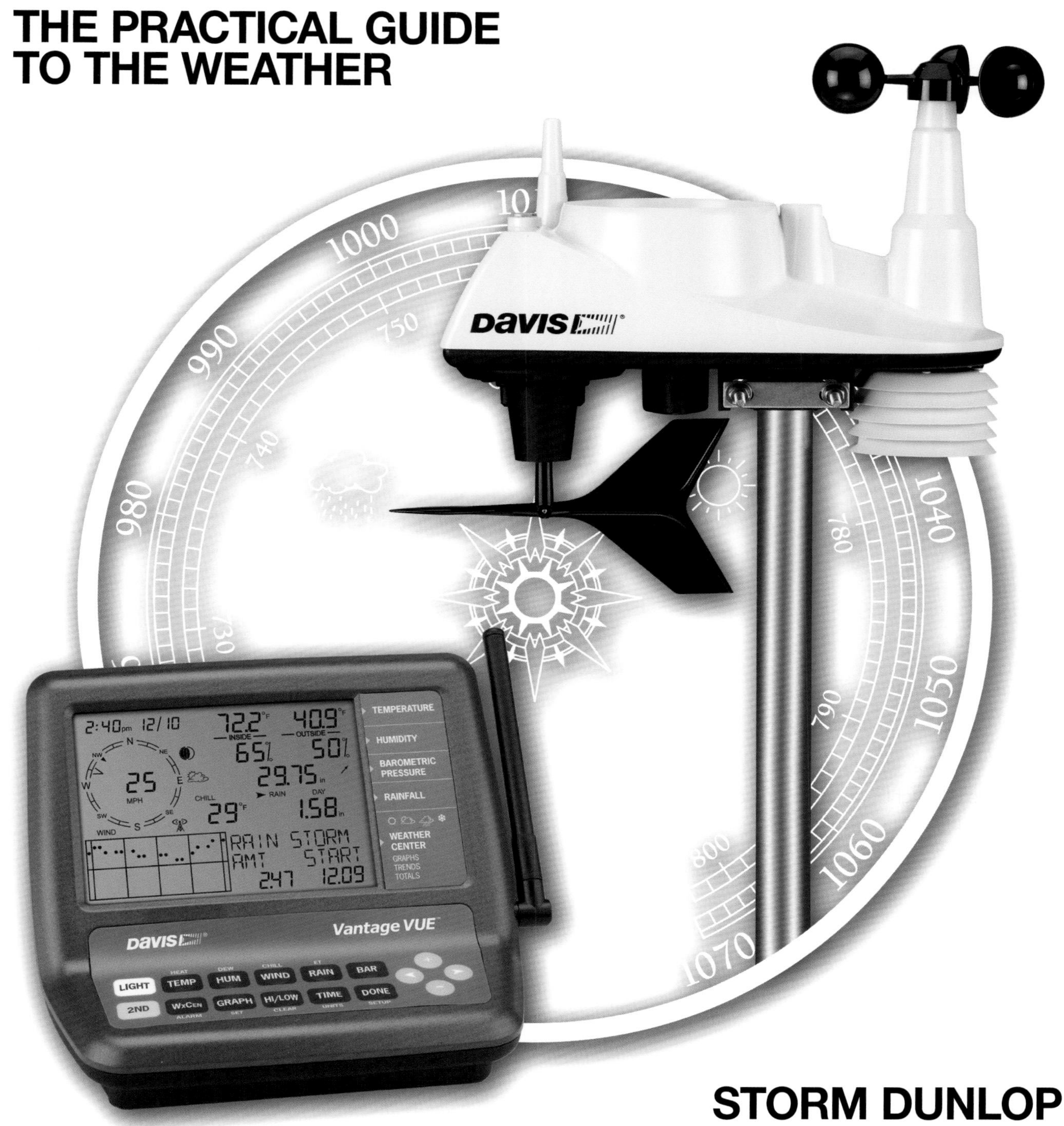

STORM DUNLOP

Contents

PART 3: OBSERVING THE WEATHER AND FORECASTING

(Author)

Introduction

Among meteorologists and climatologists there is a well-known saying: 'Climate is what you expect, weather is what you get.' In this book we are primarily concerned with weather: its causes, its observation and its forecasting. But climate – the long-term weather conditions that prevail at a particular place – strongly affects the type of weather that one is likely to encounter, so the two are closely linked. This book therefore includes some discussion of climate, but does not discuss in great detail the issues of global warming and climate change.

Accurate forecasting of the weather is an immensely complex task, and although forecasts (and forecasters) are often harshly criticised, the reliability of forecasts has improved very greatly in recent years. It is not generally recognised that forecasting the weather for a few days ahead over a relatively restricted area (such as the British Isles), requires a detailed knowledge of the conditions that prevail over the whole globe. This is perhaps more readily appreciated when it is pointed out that to produce a forecast for the weather the following day over Britain, it is essential to have information about weather systems that are on the other side of the Atlantic, some 4,800km (3,000 miles) away. Similarly, observations are required not only for conditions at the surface, but also for those at various altitudes in the atmosphere.

OPPOSITE A satellite image of Hurricane Katrina, when at its maximum strength (Category 5) on 28 August 2005. *(NASA/GSF)*

Temperatures

The majority of temperatures given in this book use metric units (*ie* degrees Celsius). However, occasionally the Fahrenheit equivalents are also shown. Note that actual temperatures – such as those measured by a thermometer – are specified by the use of the degree sign, for example '100°C'), but that differences in temperature are shown as 'deg.C' (or 'deg.F'), *eg* '-6 deg.C'.

BELOW A well-developed Cumulonimbus incus (anvil) cloud that has expanded at the tropopause, with an older anvil partially visible in the distance. *(Claudia Hinz)*

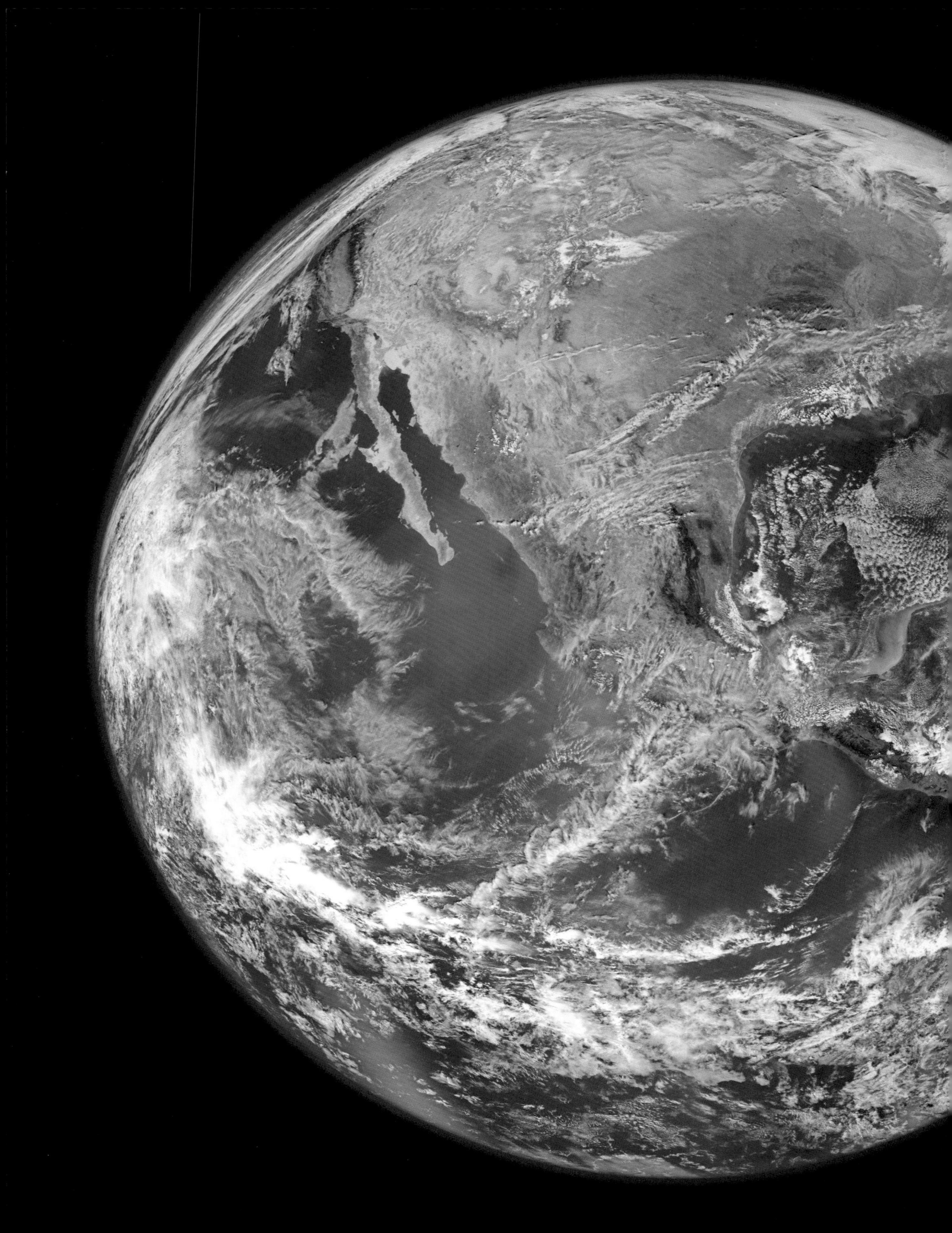

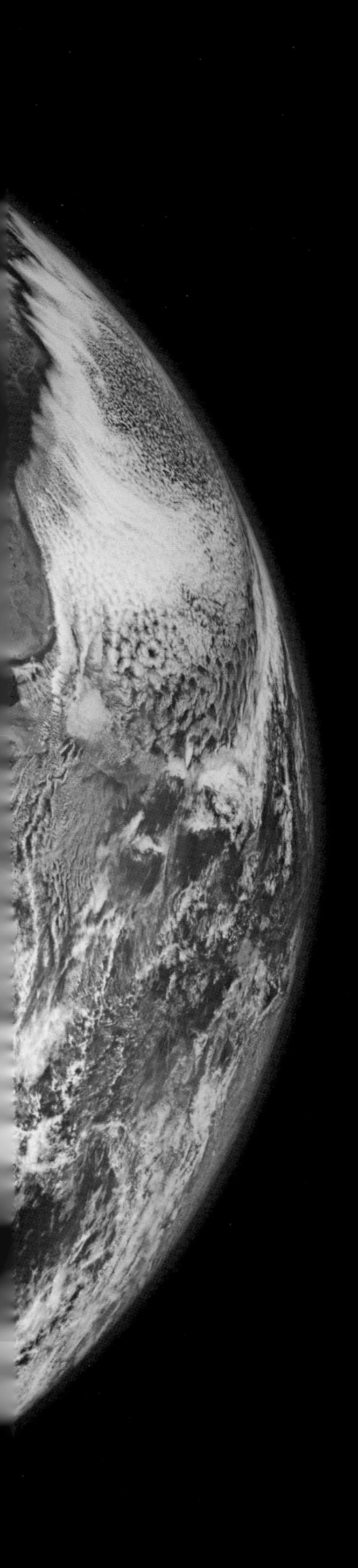

PART ONE

The atmosphere

OPPOSITE This 'Blue Marble' image of the Earth was produced by combining data from several passes of NASA's Suomi-NPP polar orbiting satellite on 4 January 2012. Although free from any dramatic features, the complexity of the cloud patterns hints at the many factors governing weather at the surface. *(NASA/NOAA/GSFC/Suomi NPP/VIIRS/Norman Kuring)*

Chapter 1

The layers of the atmosphere

The Earth's atmosphere is divided into various layers, which are most commonly defined by the way in which temperature changes with height. At first sight, the temperature might be expected to decrease from the warm surface to the cold of interplanetary space, but the actual situation is not quite so straightforward.

Meteorologists recognise five distinct layers in the atmosphere. Starting from the surface, these, with their approximate extents, are shown in the table below, together with an additional division, the ionosphere:

Atmospheric layer	Base	Top
Troposphere	Surface	15–18km to 8km
Stratosphere	15–18km to 8km	50km
Mesosphere	50km	80–95km to 100–120km
Thermosphere	80–95km	200km to 700km
Exosphere	above 200km to 700km	Interplanetary space
Ionosphere	60–70km	1,000km or more

Cross references

Cloud classification p.54
Cloud levels p.58

Most weather phenomena are confined to the troposphere (the lowest layer), together with the very lowermost region of the stratosphere. The highest clouds – which have no effect on weather at the surface – are found at the top of the mesosphere. Still higher, polar aurorae occur in the thermosphere and in the outermost layer, the exosphere. There are also other terms that are used for major layers in the atmosphere, based on specific features or composition at a particular altitude. Although not directly connected with the weather, the most significant of these divisions is the ionosphere, which includes portions of the mesosphere and thermosphere. It is itself subdivided into specific, named layers (such as the D, E, and F layers) by ionospheric researchers.

The boundaries between the various layers (where distinct alterations occur in the way in which temperature changes with height) have names that are derived from that of the underlying layer. (The mesopause, for example, is the top of the mesosphere.) The actual altitudes vary to a certain extent because of specific meteorological conditions, with latitude and with the season of the year. As we shall see in discussing the global circulation in the atmosphere, the tropopause, in particular, varies greatly in altitude between the equator and the poles, with distinct breaks at certain latitudes. The generally accepted altitudes of the atmospheric boundaries are shown in the accompanying table.

Atmospheric boundaries

Tropopause	15–18km (equatorial regions) to ~8km (poles)
Stratopause	50km
Mesopause	80–95km (winter) to 100–120km (summer)
Thermopause	200km to 700km

THE DESCRIPTION OF ALTITUDES

Although scientists generally describe altitudes in the atmosphere (such as those of the various layers and their boundaries) in metres or kilometres, the aviation industry has standardised on the use of feet. Worldwide, aircraft flight heights are always given in feet, and because of the obvious relevance of clouds to flight, cloud heights are therefore often quoted in feet. Where appropriate, equivalents are given later in this book. The World Meteorological Organization (WMO) classifies clouds in three levels (known as étages), and these are specifically discussed later.

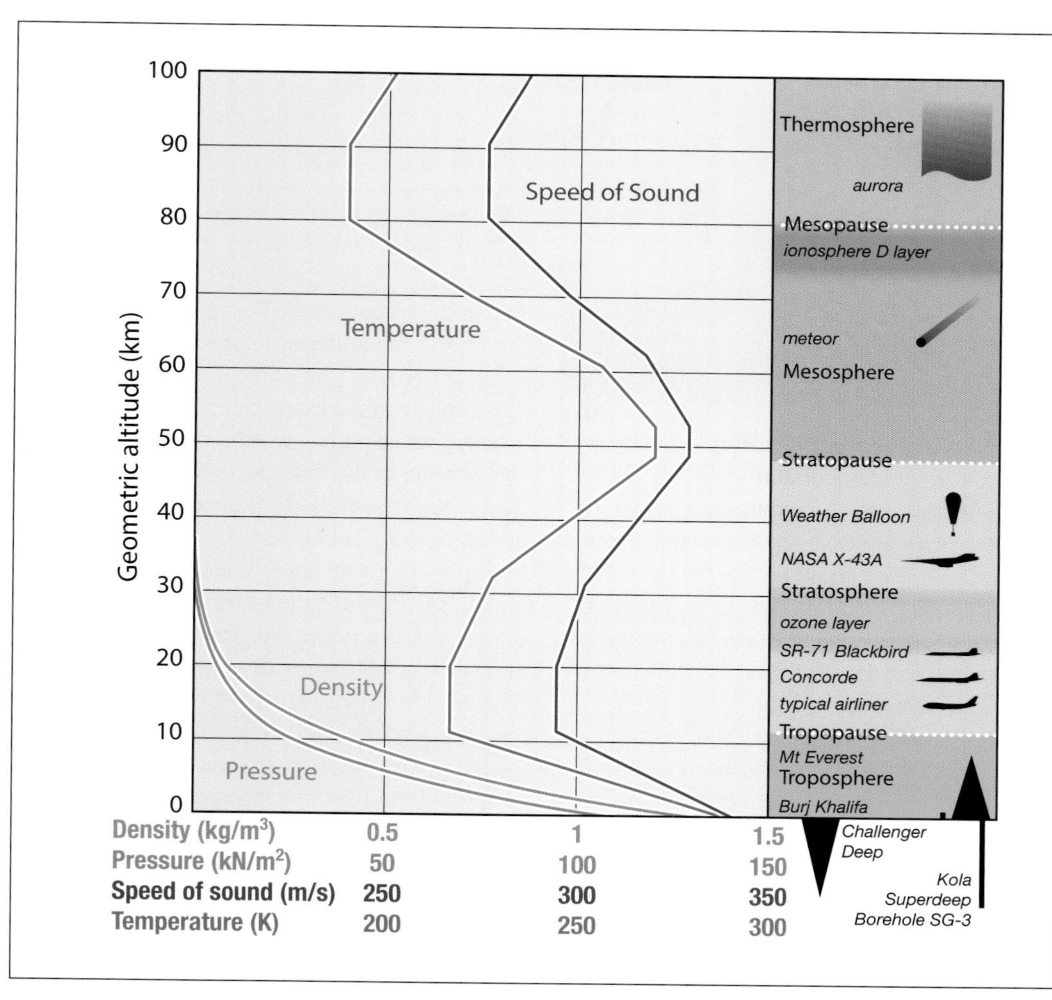

ABOVE The overall structure of the atmosphere, showing the various layers and boundaries, together with the temperature profile. *(Ian Moores)*

RIGHT Diagram of atmospheric layers. *(Ian Moores)*

The troposphere

The lowest layer is known as the troposphere (Greek for 'sphere of change'), and is where most weather occurs. When compared with the radius of the Earth (6,378km at the equator and 6,357km at the poles – 3,963 and 3,950 miles), it is extremely thin, reaching an altitude of no more than about 15–18km (~9–11 miles) over the equatorial region, and about 8km (5 miles), or sometimes even less, at the poles. Within it, the air is mixed both horizontally and vertically, and there is an overall decrease in temperature with height from the surface to the tropopause at the top of the layer. This decrease in temperature occurs because of the reduction in pressure with height. (This important relationship between temperature and pressure is of fundamental importance in various different aspects of the weather, so it will be discussed in more detail later.) Technically, the change in temperature with height is known as the lapse rate, and in the troposphere it is negative overall, decreasing with increasing height at an average rate of approximately 6.5 deg.C per kilometre. A positive lapse rate would imply an increase in temperature with increasing altitude.

Cross references

Lapse rates p.51
Temperature and pressure p.15
Temperatures and temperature differences p.30

RIGHT Low winter-time Cumulonimbus anvils. Their growth is limited by the inversion at the tropopause at an altitude of about 7 km. *(Author)*

The tropopause reaches extreme values of altitude over the equator (the equatorial tropopause, lying at 15–18km) and over the poles (the polar tropopause at an altitude of ~8km or less), with transition heights over the middle latitudes. It commonly has two distinct breaks at middle and higher latitudes. These breaks allow some exchange of air between the troposphere and the stratosphere. It may also, on infrequent occasions, appear double, where it occurs at two different altitudes, which overlap over a certain limited range of latitudes.

The overall temperature drop in the troposphere, however, is by no means steady, and the temperature may remain steady or even increase (in what is known as an inversion) through a certain vertical extent, before resuming its overall decline. Such changes in the rate of decline (or temperature reversals) that do occur are very important in the formation of clouds.

The stratosphere

At a certain altitude the temperature ceases to decrease with height. The level at which this occurs is now considered to mark the top of the troposphere. The layer above this boundary (the tropopause) is known as the stratosphere.

Above the tropopause, the temperature initially remains more or less constant – in what is called an isothermal ('equal temperature') layer – and then actually increases up to an altitude of about 50km (~31 miles). This increase ultimately arises because in the upper region of the stratosphere sunlight ionises oxygen atoms, which drift down to a lower level (roughly between 15 and 30km), where chemical reactions create ozone (O_3) with the release of energy, which heats the air. The ozone in this 'ozone layer' absorbs harmful ultraviolet radiation from the Sun, protecting the surface from its damaging effects. Man-made chemicals break down the ozone, creating 'ozone holes', particularly over the Antarctic.

Because temperature increases throughout the stratosphere, the air is stable, which means that it tends to remain at a given level. Convection and turbulence are largely absent,

Cross references

Convection p.87
Jet streams p.25
Nacreous clouds p.78
Ozone holes p.80
Stability and instability p.50

RIGHT This photograph taken from the International Space Station over western Africa shows Cumulonimbus clouds reaching up to the tropopause at an altitude of about 15km. All significant weather is confined to the troposphere, the layer below that height. *(NASA)*

and clouds are rare. There are a few exceptions in the lower stratosphere, notably in the region of jet streams and giant Cumulonimbus clouds. In addition, nacreous clouds (known more technically as polar stratospheric clouds) occur in this lowermost region.

The mesosphere

At an altitude of about 50km (~31 miles), there is another boundary, the stratopause, above which the temperature once more declines with height. The overlying layer is known as the mesosphere, and its upper boundary, the mesopause, varies in height depending on the season of the year and location, being normally at an altitude of about 80–95km (~50–60 miles), but increasing to 100–120km (~60–75 miles) near the poles in summer when upwelling occurs. The atmospheric temperature minimum occurs here, but is somewhat variable, and ranges between -163 and -100°C (-261 and -148°F). The rare, and very highest clouds ever observed in the atmosphere, noctilucent clouds (NLC) and polar mesospheric clouds (PMC), occur just below the mesopause. Winds in the mesosphere show a great variability over a period of days in winter, and this is largely absent in summer. It has also been found that there are extreme variations in pressure levels (isobaric surfaces), with great variations in height over similarly short periods of time.

The thermosphere and exosphere

Above the mesopause we come to the thermosphere, where the air density is so low that atoms and molecules rarely collide with one another and thus individually obtain very high velocities, resulting in extremely high 'temperatures'. These temperatures are strongly dependent on the level of solar activity and may reach as much as 1,500°C. The air density is so low, however, that any object in this region would undergo little actual heating. The thermopause lies at the top of the thermosphere, beyond which is the exosphere, where atoms (particularly hydrogen) may gain enough velocity to escape from Earth's gravitational field and fly off into interplanetary space. The altitude of the thermopause is extremely variable, and may range from about 200 to 700km (~124–435 miles), depending on the level of solar activity.

In the upper mesosphere and in the thermosphere the absorption of solar ultraviolet and X-ray radiation causes the atoms to become ionised (hence the name 'ionosphere' for this region), creating high electrical conductivity in various layers. These both block

Cross references

Aurorae p.78
Isobaric surfaces p.15
Noctilucent clouds and polar mesospheric clouds p.82
Pressure p.15
Stability and instability p.50
Temperature and heat transfer p.24

LEFT A display of nacreous (mother-of-pearl) clouds, photographed from Yorkshire. *(Melvyn Taylor)*

ABOVE A striking panoramic image of a major auroral display. *(Terje Nesthus)*

certain incoming radio waves and reflect other wavelengths from the ground back towards the surface, enabling long-distance radio communications. The ionosphere is generally regarded as extending from about 60–70km (~37–43 miles) to 1,000km (~620 miles) or more. Polar aurorae (the Northern and Southern Lights) are generally located within this region, at altitudes of between 100 and 1,000km (~62 and 620 miles).

There are various other ways of describing the layers in the atmosphere, but in discussing the weather we will be mainly concerned with the troposphere and stratosphere.

THE COMPOSITION OF THE ATMOSPHERE

In the lowermost layers of the atmosphere (the troposphere, stratosphere and mesosphere), the composition of the air is largely constant. The main components are:

Major atmospheric components

Gas	*Abundance (% by volume)*
Nitrogen (N_2)	78.09
Oxygen (O_2)	20.95
Argon (Ar)	0.94
Carbon dioxide (CO_2)	0.03

There are trace amounts of other gases such as neon, helium, methane, krypton, hydrogen, nitrous oxide, and xenon. Water vapour is a very variable component, ranging between 0 and 4%. The changes in humidity of the air, together with the other properties of water, have extremely important consequences in many meteorological processes, and water is discussed in detail later (p.48). The increase in the carbon-dioxide content is, of course, related to the topics of the 'greenhouse effect', global warming and climate change.

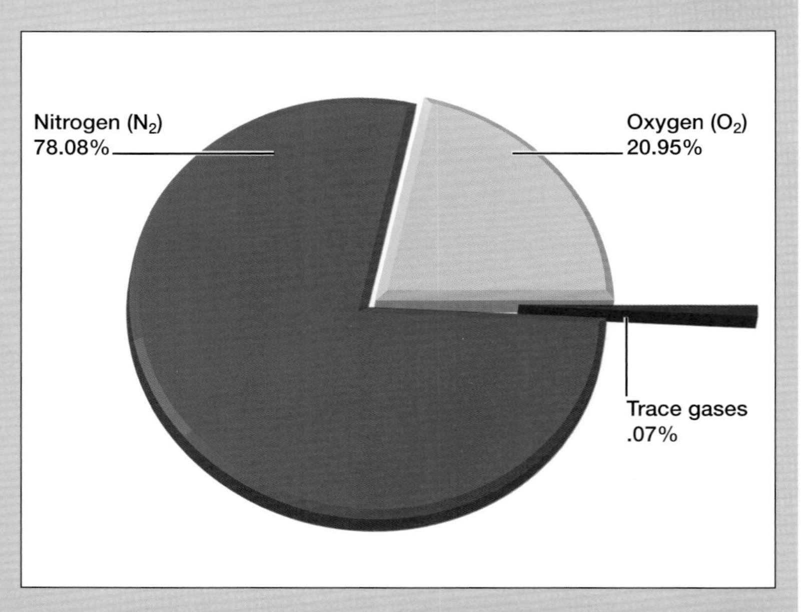

ABOVE The major components in the composition of the atmosphere. *(Ian Moores)*

Chapter 2

Pressure

Pressure is the dominant factor in determining the motion of air around the world and thus in governing current and forthcoming weather. The actual pressure at any location is determined by the weight of the column of air above that point. Unlike water, which is virtually incompressible, air is readily compressed, so the greatest pressure and density occur at the surface and decline with increasing altitude. However, both pressure and density are related to temperature. If a parcel of air – meteorologists often speak of 'parcels' or 'packets' of air – is heated, the individual molecules become agitated and increase their velocity. In a sealed container the internal pressure would rise, but in the free atmosphere the gases expand, occupying a larger volume, so the density decreases. A column of warm air will be higher than a corresponding column of cold air.

Measuring and charting pressure

Air pressure is determined with a barometer and was formerly quoted in terms of a height, often in inches, of a column of mercury. This form of measurement is still occasionally found, particularly in material from the United States. However, elsewhere in the world – and increasingly in US forecasts – it has now been generally superseded by a unit known as the millibar (mb), which is nominally one thousandth of a bar. One bar is approximately the value at sea level on Earth. However, pressures are rarely given in bars except in the discussion of the extreme pressures found in the atmospheres of the giant planets (Jupiter, Saturn, Uranus and Neptune). In fact, for standardisation purposes, the average pressure at sea level on Earth has been defined as 1,013.25mb. In most countries and forecasts, pressures are given in millibars.

Technically, pressure should be measured in the standard metric unit, the Pascal (Pa), or its standard multiples, such as kilopascals (kPa). For convenience, however, because the Pascal is rather small, meteorologists usually give pressures in terms of the hectopascal (hPa), where one hectopascal (1hPa) equals 100Pa. This has the great advantage that one hectopascal is identical to one millibar (1mb).

Surface pressure is represented on charts by isobars, lines connecting points of equal pressure, most frequently at intervals of 4hPa. Although such charts are of great value in determining the current and probable future weather, the pressures are adjusted to those that would apply at sea level, using an appropriate formula. In mountainous areas, in particular, they will differ considerably from the pressure measured at the ground. For forecasting purposes and for scientific study, charts are often produced to various pressure levels (isobaric surfaces) in the atmosphere, when the contour lines commonly represent the altitude at which the particular pressure is located. Such charts are commonly prepared for pressures of 1,000, 850, 700 and 500hPa. The difference between adjacent levels is known as the 'thickness' and is an indication of the temperature of that particular layer. Large values indicate a deep warm layer, small ones a shallow cold one. At any level (including the surface) the closer the isobars, the stronger the wind.

As mentioned, air pressure naturally decreases upwards from the surface to interplanetary space, simply because of the decrease in the pressure that the overlying air exerts on layers beneath it. From 1,013mb at sea level, pressure falls to approximately 500hPa roughly halfway to the top of the troposphere. At the tropopause, the pressure is typically about 200hPa. Under normal circumstances, ignoring extreme events such as tornadoes and tropical cyclones, sea-level pressures have a range of about 100hPa, from about 950hPa to 1,050hPa.

Chapter 3

The global circulation

The various weather systems across the world are ultimately driven by the imbalance between the amount of energy received from the Sun at low latitudes and that incident on polar regions. Far from being a simple overall overturning, a complex system of circulation cells, winds and ocean currents transports heat from equatorial regions to cooler latitudes.

The overall global circulation arises from the transport of heat from the tropics towards the poles by the system of winds. There are three major factors that determine the strength and direction of winds:

- The difference in pressure between the two points (known as the pressure gradient). This difference is normally created by temperature differences, which create air masses of different densities.
- The rotation of the Earth.
- Friction with the surface. The amount of friction obviously depends on the nature of the surface over which the air is flowing.

All of these factors have a part to play in the overall global circulation, but the effect of the Earth's rotation is only occasionally of significance in local wind systems.

Air pressure across the world

The differences in heating by the Sun that occur across the world create differences in surface air pressure. More energy is received than lost within a zone from the equator to about 40°N and S, and a there is a deficit from those latitudes to the poles. Although changes naturally occur with the seasons and from one year to the next, the overall pattern remains similar from year to year. Sea temperatures also have an effect on global weather. The shift in sea temperatures between the seasons is also noticeable, although this tends to lag behind changes in land surface temperatures.

The temperature pattern translates into a pressure pattern which then determines the overall distribution and direction of global winds. There is a low-pressure region (the equatorial trough) in the tropics, between the equator and latitudes of 5–10°N and S, where warm air is rising. There are high-pressure regions

CHANGES IN WIND DIRECTION

Two terms are used to describe changes in the direction of the wind that occur over a fairly short time (such as with the passage of a frontal system). Note that they apply in both hemispheres. These are:

- Veer – a clockwise change, *eg* from east to south. When facing the wind, its direction moves to the right.
- Back – an anticlockwise change, *eg* from west to south. When facing the wind, its direction moves to the left.

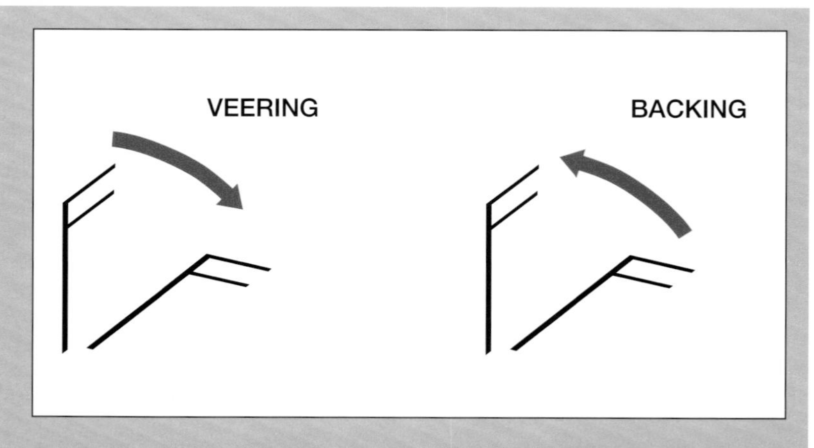

ABOVE Veering is a clockwise change in the direction of the wind, and backing an anticlockwise one. *(Ian Moores)*

RIGHT Global temperature distributions for January & July 2012. *(Ian Moores)*

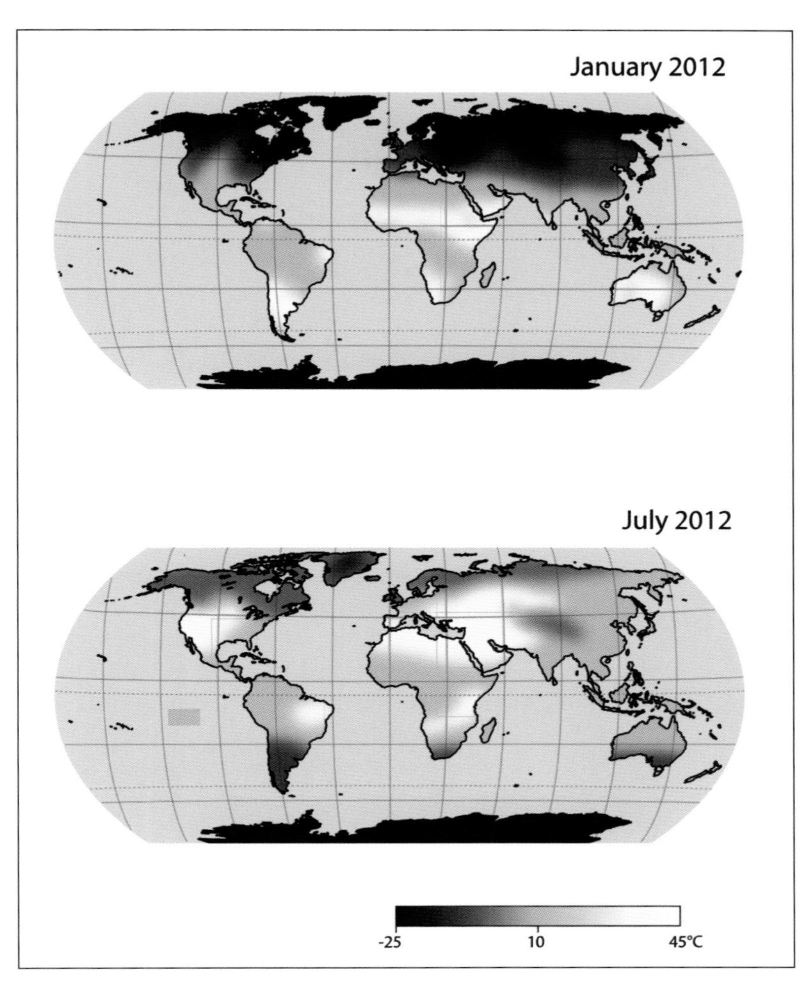

(the subtropical anticyclones) at approximately 25–30°N and S, where air is descending; temperate low-pressure zones at about 40–70°N and S; and high-pressure regions over the poles, beyond about 70°N and S.

A few pressure features are permanent or semi-permanent and consistently appear in specific regions. These are known as 'centres of action' because they have a significant effect on the development of weather systems in surrounding regions. More or less permanent examples are the Azores and Pacific Highs, and the Icelandic and Aleutian Lows. Semi-permanent features are the Canadian and Siberian Highs that develop in the northern winter, and the Asiatic Low in the northern summer. The Antarctic anticyclone is a year-round feature, whereas the Arctic cold, high-pressure region is only prominent in the northern winter.

The circulation cells

In 1686, Edmond Halley (of comet fame) gave the first serious atmospheric circulation. He pointed out that warm air rose at the equator and drew in colder air from nearby regions and thus accounted for the persistent trade winds. He published the first meteorological chart showing the winds in the tropics and adjacent latitudes. Although his explanation of the overall circulation was at fault, his suggestion that the circulation was driven by temperature differences is correct, so he is still occasionally termed the 'Father of Dynamic Meteorology'. A meridional (north–south) circulation like that proposed by Halley would occur on a non-rotating planet and a similar circulation does occur on Venus, which has an extremely slow rotation period of about 243 days.

In 1735, George Hadley suggested that there was a large circulation cell in each hemisphere,

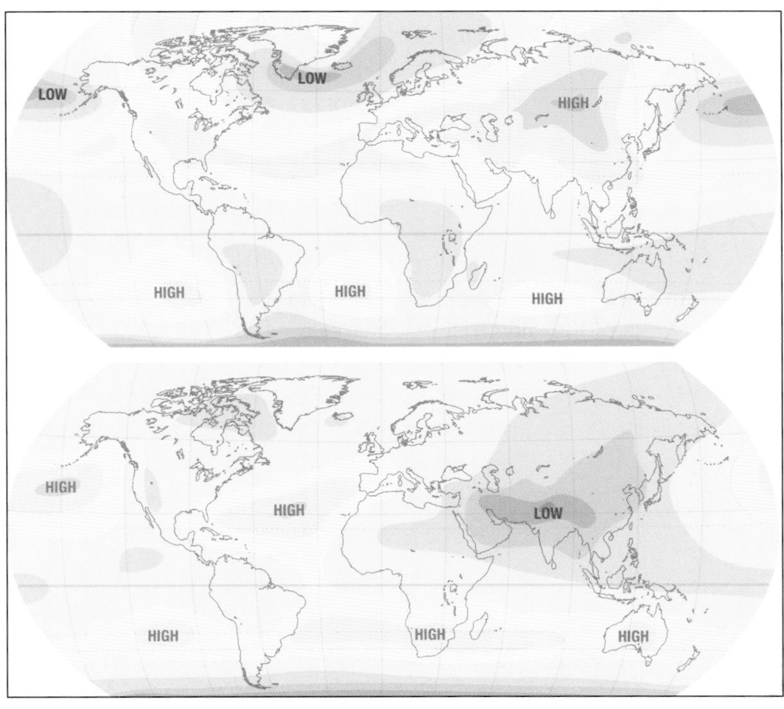

RIGHT Average global pressure distributions for the months of January (top) and July (bottom). *(Dominic Stickland)*

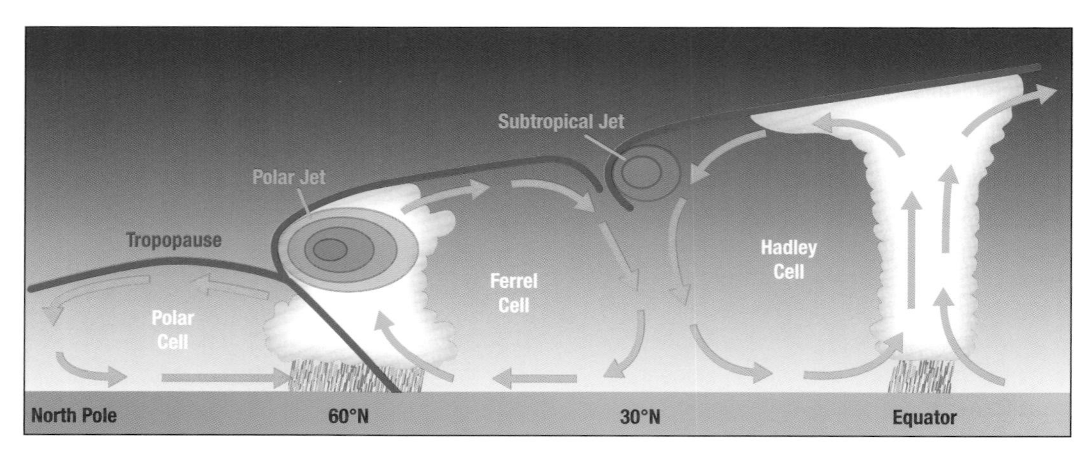

RIGHT A schematic representation of a cross-section of the three northern atmospheric circulation cells. *(Dominic Stickland)*

where warm air rose above the equator, flowed at height towards the poles, and reached the middle latitudes, where it cooled and sank. It then flowed back towards the equator at low level, forming the trade winds. Although this concept of a single meridional circulation cell was incorrect, Hadley correctly explained that the directions of the north-easterly and south-easterly trade winds converging on the equator were diverted from a direct north–south (and south–north) direction because of the rotation of the Earth. However, he was unable to account for the band of persistent westerlies that are found in the middle latitudes. He thus explained the meridional circulation, *ie* the circulation in a north–south and south–north direction, along lines of longitude, but not the strong zonal (latitudinal) circulation.

It was eventually established that there are three primary circulation cells in each hemisphere. In one, warm moist air rises over the equatorial zone and descends at the middle latitudes, around 25–30° north and south. This particular circulation is known as the Hadley Cell, in recognition of Hadley's contribution. It is basically a closed loop, driven by convection, with the air that descends at 25–30° N and S returning towards the equator as the trade winds. This cell is particularly deep over the equatorial zone where the tropopause may be as high as 15–18km.

The boundary where the north-easterly and south-easterly trades meet is known as the Inter-Tropical Convergence Zone (ITCZ), often clearly visible on satellite images as a line of clouds extending approximately parallel with the equator.

Another cell, the Polar Cell, exists in the polar regions where cold dense air spreads out at the surface and moves towards lower latitudes. There is still a certain amount of heating by the Sun even at latitudes around 60°N and S, so here again, a weak convective, closed-loop cell is created, although this is much shallower, being limited by the low polar tropopause to a depth of about 8km.

Both the Hadley Cell and the Polar Cell are thermally direct – that is, they are both driven by temperature differences.

Between these two cells lies a third, the Ferrel Cell (named after the meteorologist who first postulated its existence, William Ferrel). Unlike the other cells it is thermally indirect. In other words, its circulation is primarily driven by the circulation of the other two cells. Its motions are required to complete the overall global circulation. Not all of the air that descends at 25–30°N and S returns towards the equator as the trade winds. Some spreads out towards the poles until it encounters the circulation in the Polar Cell. Here it rises from the surface, with most of the air flowing back towards 25–30°N and S at height but some is entrained into the circulation of the Polar Cell. The region where the Polar and Ferrel Cells meet is an extremely important atmospheric boundary, known as the Polar Front, which has major effects on the development and movement of weather systems.

Although the general circulation was initially believed to be a meridional circulation, with flow taking place in an essentially north–south plane, in none of these three cells does the air actually flow directly north or south. There is always a substantial zonal (or latitudinal) flow. This is a result of the rotation of the Earth: the factor that Halley omitted and which Hadley incorrectly interpreted.

Cross references

Jet streams p.25
Weather systems p.39

There are also very significant zonal flows in the jet streams, the extremely strong wind systems in both hemispheres, that blow just below the tropopause near the boundaries of the cells, approximately at latitudes 30° (the Subtropical Jet) and 60° (the Polar Jet). The Polar Jet, in particular, greatly influences the development and motion of pressure systems at the surface.

The Coriolis effect

The Earth's rotation affects the direction of winds and ocean currents (as well as other moving objects such as artillery shells and missiles) through what is known as the Coriolis effect. This applies to any moving object that is observed in a rotating reference frame (such as that experienced by an observer on the rotating Earth). The effect was first formulated mathematically by the French scientist Gaspard-Gustave Coriolis, after whom it was subsequently named, although it had been described earlier by others.

The result of the effect is that a body may move in a straight line, but appears to follow a curved path. Although the effect occurs in three dimensions, it is most commonly evident in the motion of air in winds and pressure systems in two dimensions relative to the surface of the Earth. This horizontal component is often known as the Coriolis force. It may be understood, in simple terms, by considering parcels of air at the equator and at one pole. At the equator, a stationary parcel is being carried eastwards by the rotation of the Earth at a rate of 40,074km (24,900 miles) in 24 hours and thus at a velocity of approximately 1,670kph (~464m/s or ~1,038mph), whereas a parcel of air at one of the poles has no eastward motion at all, but merely rotates around its axis. To take the case in the northern hemisphere, if the equatorial parcel of air is subject to some pressure gradient that moves it north, it retains its high eastwards velocity but, away from the equator, the surface beneath it is moving slower. The parcel will move east, relative to the lines of longitude. From the point of view of an observer on Earth, the air has curved to the right. If the parcel moves south into the southern hemisphere, exactly the same effect will occur, except that it will appear to move to the left.

At one of the poles the eastwards motion is non-existent. (Let us take the North Pole.) If the air starts to move towards the equator, it has no eastward velocity, but the surface now beneath it is moving faster towards the east. Again, the parcel will appear to curve towards the right. The situation at the South Pole is similar, but here the curvature will be to the left. Note that this description is highly simplified. In reality the effect arises because of the spin associated with any parcel of air, located anywhere on Earth. (This spin is a maximum at the poles, and a minimum – non-existent – at the equator.) Most importantly, the effect occurs regardless of the direction of motion of the parcel of air – whether the motion is meridional (north–south) or zonal (along a line of latitude).

In other terms, in the northern hemisphere the wind veers (turns clockwise) and in the southern it backs (turns anticlockwise).

The Coriolis force is least at the equator. The magnitude is actually proportional to the sine of the latitude, so its magnitude is zero at the equator (sin 0° = 0), and reaches a maximum at the poles (sin 90° = 1). It also varies with the horizontal velocity of the air (the greater the

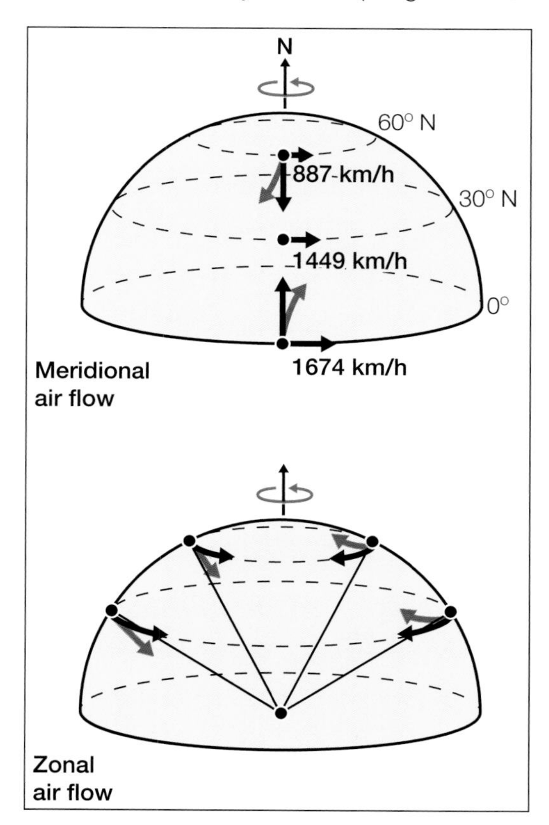

LEFT The Coriolis effect as it applies to meridional airflow (top) and zonal flow (bottom). *(Ian Moores)*

Cross references

Depressions p.36
Stratosphere p.12
Tornadoes p.121
Tropical cyclones p.124

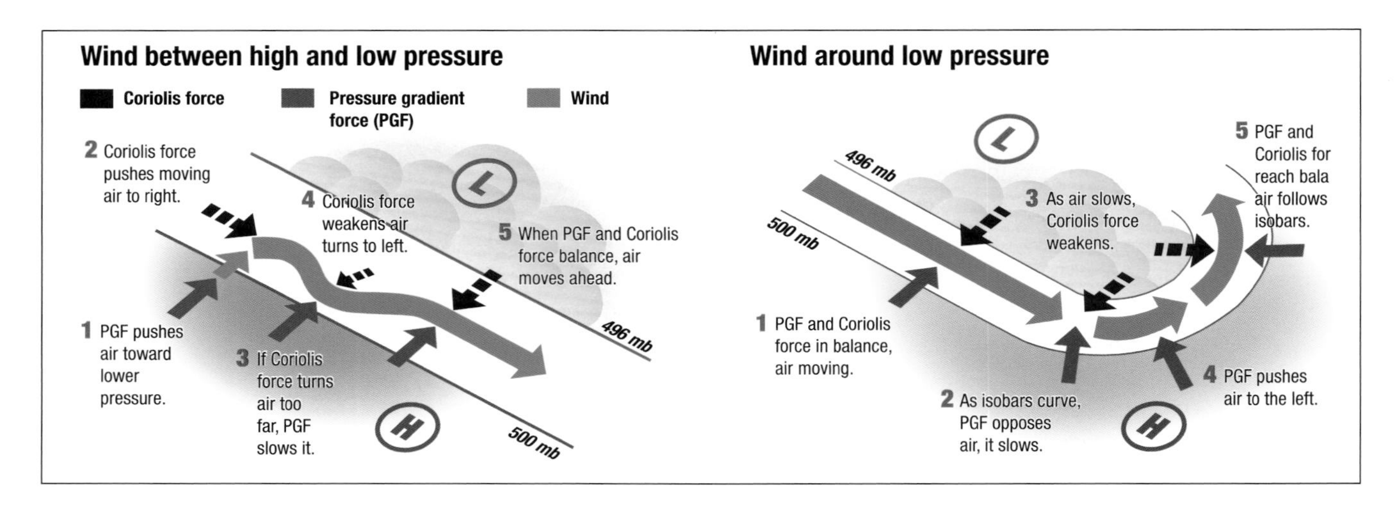

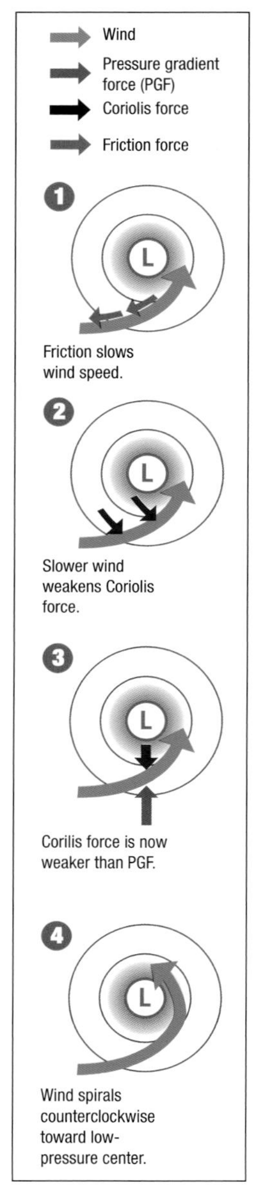

ABOVE AND RIGHT **How air flows between high and low pressure at altitude around low- and high-pressure systems when the air is able to move freely.** *(Dominic Stickland)*

LEFT **How friction with the surface modifies the airflow, causing it to spiral in towards the centre of the low.** *(Dominic Stickland)*

BELOW **How the gradient wind at jet-stream level flows past a low-pressure region.** *(Dominic Stickland)*

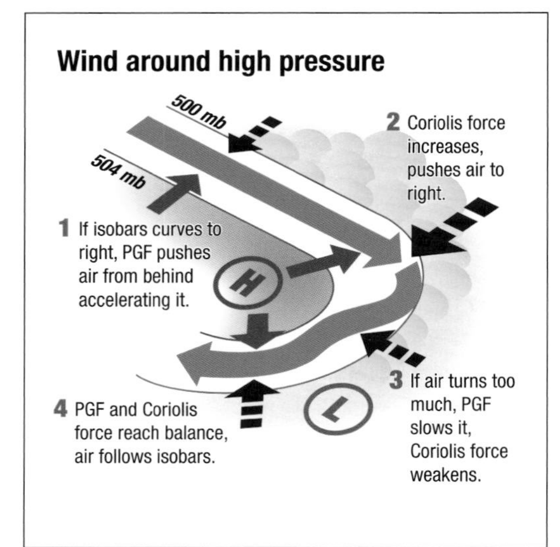

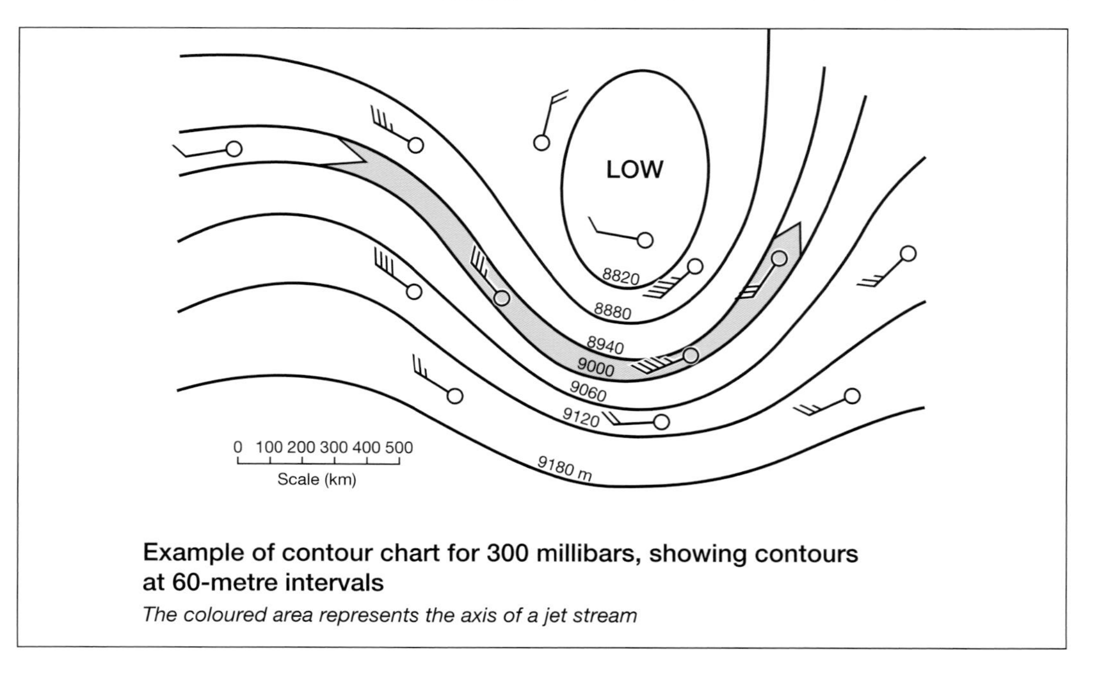

Example of contour chart for 300 millibars, showing contours at 60-metre intervals

The coloured area represents the axis of a jet stream

speed, the greater the force), and always acts at right angles to the direction of motion. This is one reason why tropical cyclones (hurricanes, typhoons etc) do not develop until the initial system moves away from the equator.

The Coriolis effect is minute, and generally has an effect only on systems where it acts over a long distance or for a long time. In meteorology, the effect is important in large-scale weather systems (such as a depression that may be several hundred kilometres across), but is negligible on a small scale. It plays no significant part, for example, in a tornado that is perhaps a kilometre in diameter, where the pressure gradient and other rotational forces are far more significant.

There is one very important, and initially surprising, result of the Coriolis effect. The pressure gradient acts to force air across the isobars, from high pressure to low, but the Coriolis force immediately comes into effect, causing the air to turn to the right (in the northern hemisphere). The two forces eventually balance one another exactly. This causes the air to flow along the isobars, rather than across them. This is the prevailing state in the free atmosphere, that is, away from the influence of the surface. This freely flowing wind is known as the geostrophic wind, and generally occurs at altitudes greater than approximately 500m over the oceans and 1,500m over land.

At low levels, friction with the surface comes into play, slowing the wind. Because the Coriolis force is proportional to the wind speed, it weakens relative to the pressure-gradient force, so the Coriolis curvature decreases. The pressure gradient becomes relatively stronger and the air begins to flow across the isobars. The result (in the northern hemisphere) is that the air flows anticlockwise into the centre of low-pressure areas, and clockwise out of high-pressure regions (and in the opposite directions in the southern hemisphere). The extent of the curvature depends on the amount of friction experienced by the flow of air, less over the oceans and more over land. The difference in direction may amount to about 10–15° over the sea, and as much as 40–50° over the land. These differences are seen in localised effects. For example, where a wind with a long fetch over the sea strikes a coastline, the additional friction may cause the wind inland to slow and back considerably. Similarly, an offshore wind may accelerate and veer when the airflow passes out to sea.

Locating the general direction of the centre of low pressure may be determined by Buys Ballot's Law (named after the Dutch meteorologist who first formulated it, C.H.D. Buys Ballot): 'Stand with your back to the wind, and the low is on the left.' (In the northern hemisphere, of course.) This is only an approximate indication because, rather than lying directly to the left, the actual centre lies farther forward as a result of the effects of friction.

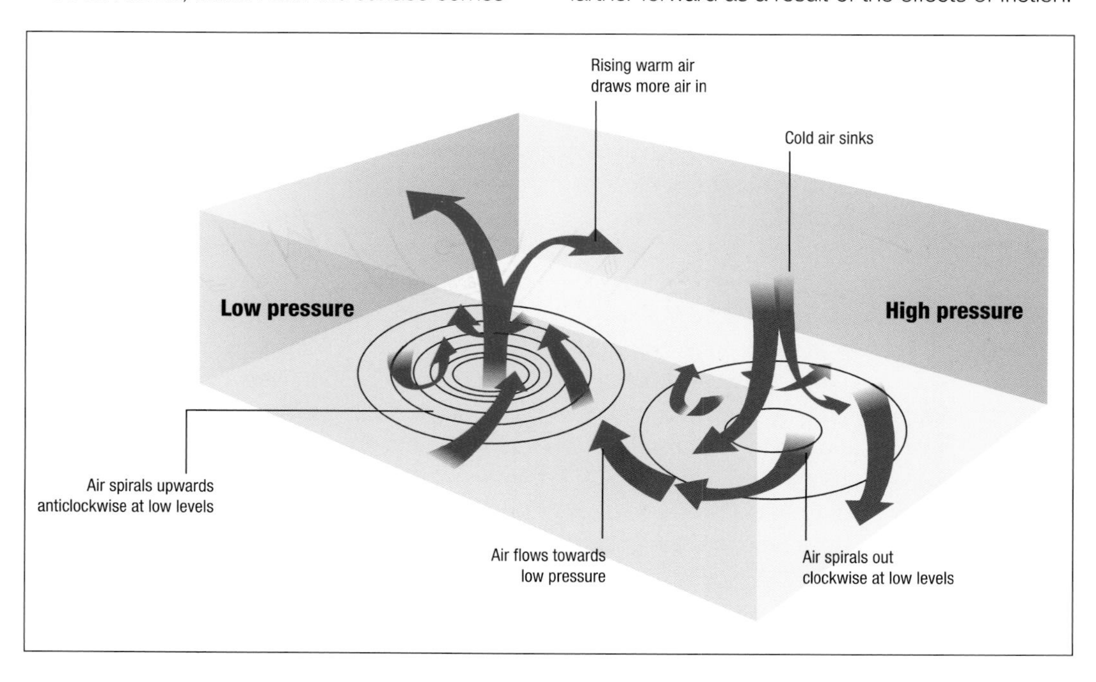

LEFT A schematic diagram of how air flows into, through, and out of low- and high-pressure systems. *(Dominic Stickland)*

RIGHT Typical global wind patterns for January and July. *(Dominic Stickland)*

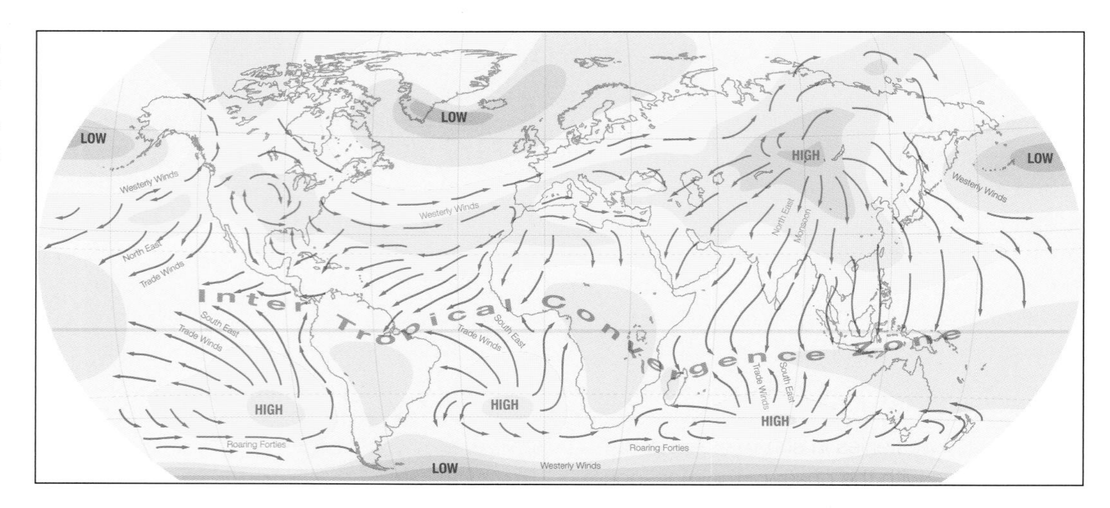

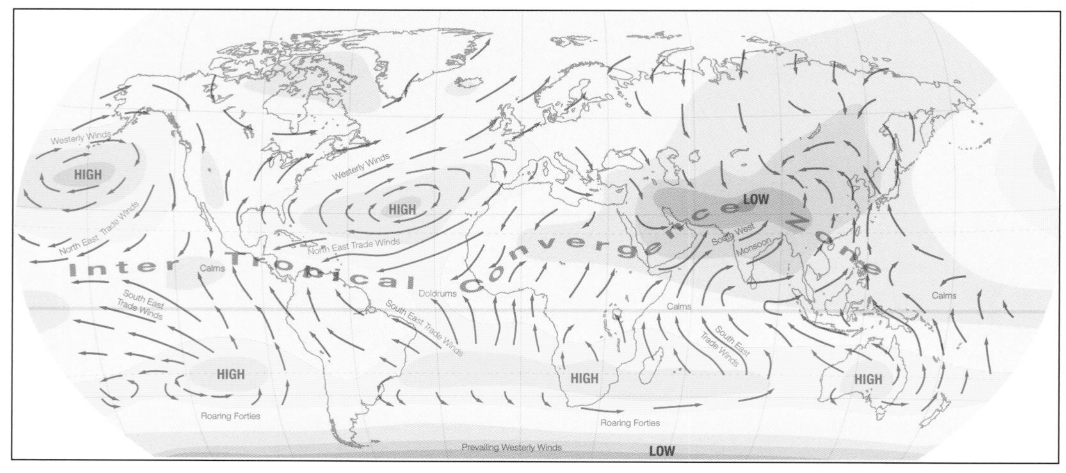

The Coriolis effect and the global circulation

One result of the Coriolis effect is that the air circulating within the global cells does not flow directly north and south. In the northern hemisphere, the air, rising at the equator in the Hadley Cell, flows approximately north-east at height, descends at around 25–30°N, and then flows back towards the equator at the surface as the north-easterly trades. Where the trade winds converge (at the Inter-Tropical Convergence Zone) in the equatorial trough, the winds over the oceans tend to be light and variable in direction, and this region became known as the Doldrums.

In the northern Polar Cell, the dense, cold air that flows out of the Arctic produces the polar easterlies. In the intermediate Ferrel Cell, air that descends at 25–30°N and flows towards the pole, curves towards the east and contributes to the strong, prevailing westerlies that are so dominant in the middle latitudes between 30 and 60°N. A similar situation applies in the southern hemisphere, and here the westerlies are even stronger and more persistent because there are few continental land masses to interfere with their flow around the globe. This gives rise to the band of strong, persistent winds known as the Roaring Forties and, even further south, to those known informally as the 'Fearsome Fifties' and the 'Screaming Sixties'. These are bounded on the south by the Antarctic Front, where the extremely cold, dense south-easterlies flow down and out from Antarctica itself. Often lying between approximately 60 and 65°S, the atmospheric boundary marked by the Antarctic Front may, at times, extend almost completely round the globe.

The simplistic pattern of winds just described is considerably distorted over the Indian and

Cross references

Jet streams p.25
Polar vortex p.80–81

western Pacific Oceans during the northern summer. This is because of the development of the Asiatic Low, a thermal low produced by intense heating of the continental interior. This draws in air from surrounding regions to such an extent that the Inter-Tropical Convergence Zone is indistinct and air flows across the equator into the continental interior. It is this reversal of the winds that gives rise to the monsoon regime over India and surrounding countries, with a generally dry, north-easterly airflow during the winter that reverses to warm, humid south-westerlies in summer that bring torrential rains with the onset of the monsoon. The monsoon-wind reversal is particularly strong over the north-western Indian Ocean. Less energetic monsoon systems exist elsewhere in the world, such as over the south-western United States.

The changes in the rainfall regime in Asia is the most dramatic effect of the alterations in the pattern of prevailing winds between the seasons. The precipitation along the ITCZ moves north and south but, in particular, rainfall over Asia switches from generally dry during the northern winter (when the Siberian High is dominant) to wet during the summer.

Although most of the world's prevailing winds are primarily driven by the latitudinal difference in solar heating, certain significant longitudinal circulations occur, often driven by the fact that water absorbs more heat than air, and releases it more slowly. (This is significant in some localised winds discussed later.) Temperature contrasts tend to strengthen some of the existing wind patterns, particularly true over the Pacific, where the normal convergence of the trade winds at the ITCZ is intensified into easterly winds along the equator.

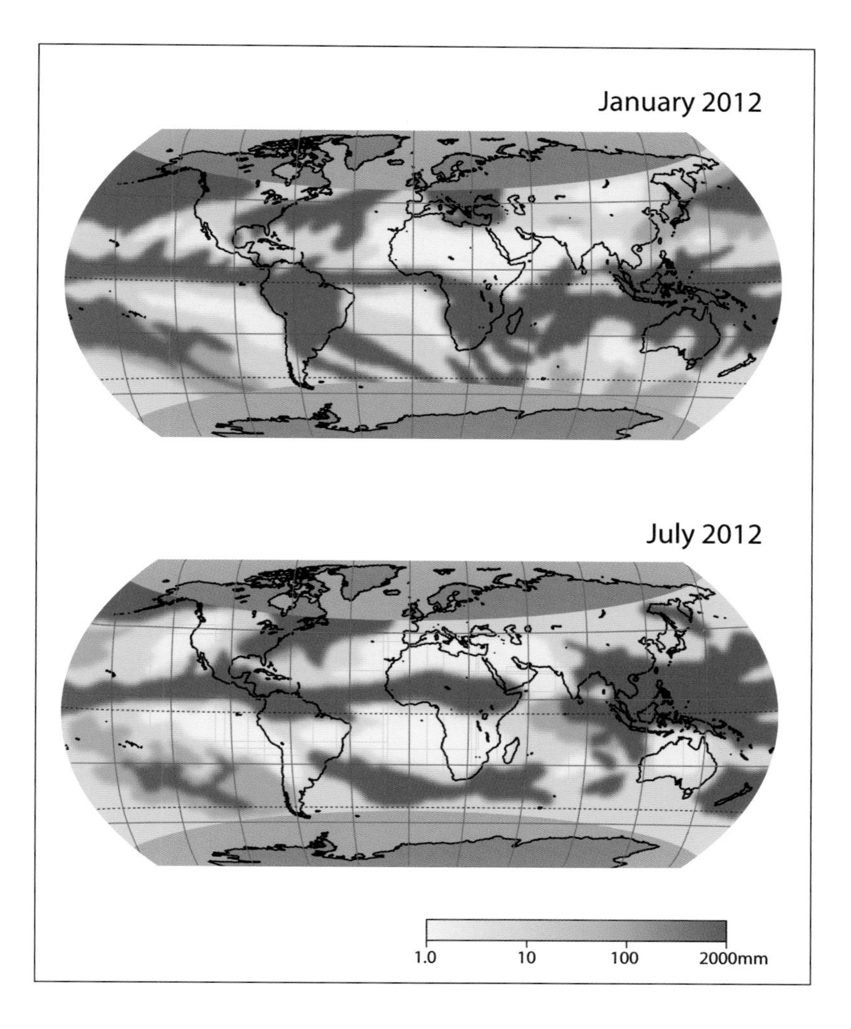

ABOVE Maps of average overall global precipitation for January and July. *(Ian Moores)*

Oceanic currents

The global pattern of winds acts on the upper layer of the oceans, giving rise to the surface currents. These currents also act to distribute heat from the tropics towards the polar regions. Because of the oceanic Ekman effect the directions of the currents deviate slightly from the actual wind directions.

There is a major circulation (known as a gyre) in each of the main oceanic basins, clockwise in the northern hemisphere and anticlockwise in the southern. These gyres are asymmetrical, with the western currents, flowing towards the poles (known as the western boundary currents), being narrower and faster than the southward-flowing eastern boundary currents. The western currents transport large amounts of heat towards the poles, much of which is transferred to the air above them. The eastern boundary currents are generally broader, much cooler, and promote upwelling of bottom water (usually accompanied by major fisheries). There are some changes in the rates of flow and latitudinal displacement of the ocean currents with the seasons, although this mainly affects those portions of the gyres that flow west,

Cross references

The Coriolis effect p.19
Global wind patterns p.22

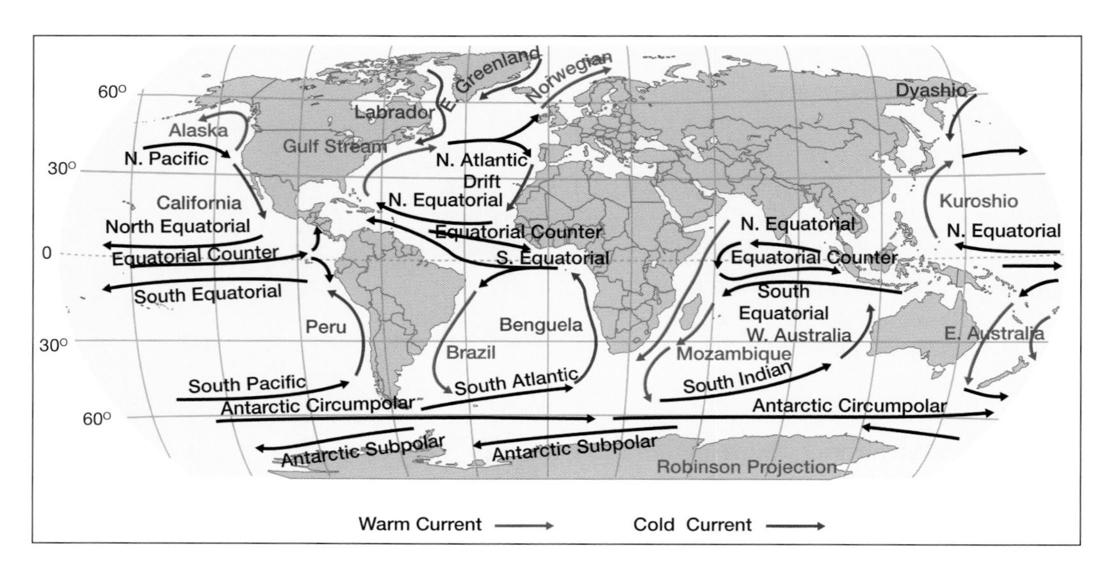

RIGHT The major oceanic currents. *(Ian Moores)*

approximately parallel to the equator. In the equatorial regions there are also surface counter-currents flowing towards the east, between the main flow in the gyres at those latitudes.

Climate

The combination of the heat transferred by the ocean currents and by the atmosphere gives rise to the various climatic regimes around the world. The climate at any particular location is primarily determined by the latitude, its situation relative to centres of action, its proximity to the oceans (oceanity or continentality), its altitude, and also to the local topography (whether it is affected by neighbouring mountain ranges and other factors). There are various climate classification schemes, but one that is widely used is the Köppen classification (sometimes called the Köppen-Geiger system), which is based on the annual and monthly ranges of temperature, the average rainfall (particularly the seasonal variations), the wind regime, and the overall type of vegetation. The complexities of the classification system need not concern us here, but there are two very broad climate types: continental and maritime climates.

The air that subsides in the subtropical anticyclones (the significant centres of action at latitudes 25–30°N, mentioned earlier) warms, and becomes less humid as it does so. The world's major deserts, particularly the Sahara and Arabian deserts and the smaller desert area in the American south-west, lie beneath this region of subsiding air. In the southern hemisphere, the subtropical anticyclones' influence is more limited, because of the smaller continental areas, but they help to create the conditions in the Kalahari in southern Africa and in the central Australian deserts.

Other desert areas arise because of the proximity of cold ocean currents and upwelling. The cold water leads to a lack of evaporation into the air above, so that even when winds are off the ocean, neighbouring coastlines experience little rainfall. This is particularly the case in the southern hemisphere, where the driest place on Earth, the Atacama Desert, receives minimal precipitation from the adjoining Pacific, where the Humboldt Current and its associated upwelling of cold bottom water are particularly strong. A similar effect occurs with the Benguela Current in southern Africa, leading to the dry conditions in the Namib Desert. In the North Atlantic, the cold Canary Current produces little precipitation over north-west Africa, already under the influence of the descending air over the Sahara.

Some other desert areas, especially the Gobi Desert and surrounding regions in the Asian heartland, obtain little precipitation simply because they are so far from the sea that most of the airstreams have lost their humidity before they penetrate into the interior. In addition, technically, both the Arctic and, especially, the Antarctic are 'deserts' because precipitation is exceptionally low. (The vast East Antarctic ice sheet has accumulated over millions of years,

Cross reference

Centres of action p.17

despite the minute amounts of snow that fall in the interior in any one year.)

By contrast, warm ocean currents transfer heat to the air above them, and thus cause neighbouring coasts to be warmer and more humid than might otherwise be expected for their latitude. The warm Alaskan Current, which circles in an anticlockwise direction round the Gulf of Alaska, produces a maritime climate in north-western North America (the US states of Oregon, Washington and Alaska, and British Colombia in Canada).

In the southern hemisphere, warm currents produce mild, moist climates (and the accompanying weather) on the eastern coasts of South America (Brazil and Argentina), southern Africa (Mozambique and South Africa), and Australia (where the mountainous Great Dividing Range creates a strong contrast between conditions on the coast and those prevailing in the dry interior to the west).

Jet streams

Although there is a general zonal flow around the Earth, within it there are jet streams: narrow, fast-flowing ribbons of air, the most important of which lie near the tropopause. They may be thousands of kilometres long, hundreds of kilometres wide, and just a few kilometres deep. The definition of a jet stream is when it has a velocity of more than 25–30m per second (90–108kph or 56–67mph). Jets arise, like winds closer to the surface of the Earth, through strong contrasts in temperature, and the resulting differences in density and thus pressure. Significant jets tend to be located at the boundaries of the major circulation cells. The resulting pressure gradients, together with the Coriolis effect, create strong westerly jets.

LEFT The northern Polar Front jet stream over Nova Scotia, flowing eastwards above the Atlantic, photographed from orbit. *(NASA)*

In each hemisphere there are two major, westerly jet streams: the first, the Polar Jet, lies at altitudes of about 7–12km (23,000–39,000ft) and between latitudes 40 and 70°N and S. Because of the slope of the Polar Front, the jet generally lies on the polar side of the surface front. The second, the Subtropical Jet, is both weaker and higher, at an altitude of about 10–16km (33,000–52,000ft). (Unlike the Polar Jet, it is not associated with surface frontal systems.) The approximate locations of these are shown in the diagrams. The strength of jet streams is, of course, strongly dependent on the temperature contrasts that exist. For this reason, the Polar Jets tend to be more powerful

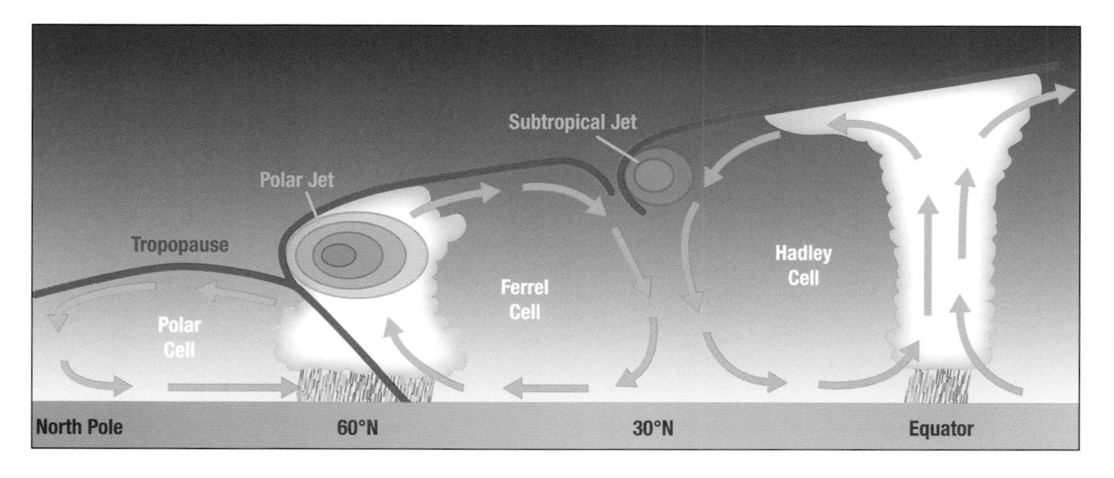

LEFT The three circulation cells and the two major jet-stream systems in the northern hemisphere. *(Dominic Stickland)*

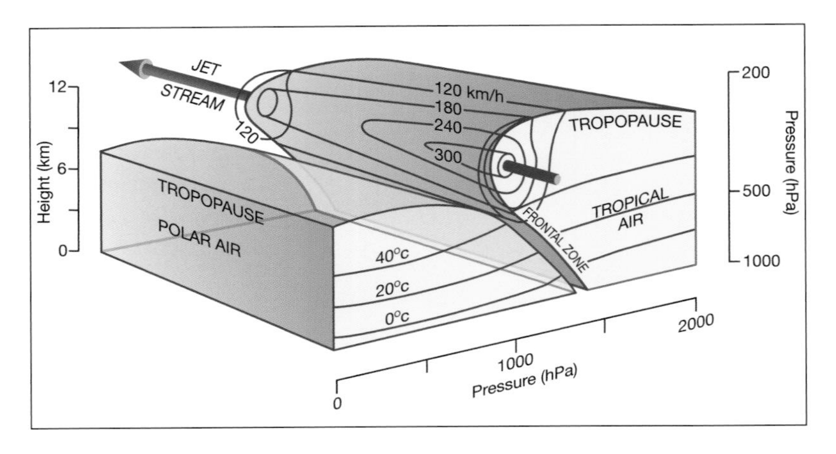

ABOVE A three-dimensional representation of the frontal zone (at the northern Polar Front), showing the different heights of the tropopause on the two sides of the front and the location of the strong Polar Jet. *(Ian Moores)*

ABOVE A schematic representation of the locations of the main jet streams, with a polar and a Subtropical Jet in each hemisphere. *(Dominic Stickland)*

than the Subtropical Jets. Similarly, the strength of the Polar Jets (in particular) increases during the winter season, when the polar air becomes especially cold, because of the absence of sunlight and solar heating within the Arctic or Antarctic Circles.

Other jet streams also exist. An easterly tropical jet (the Equatorial Jet Stream) sometimes forms in the northern summer over the eastern hemisphere at a latitude of approximately 10°N and an altitude of 15–20km, where there are extreme temperature contrasts, and when the coldest air actually lies over the equatorial region. This jet does not extend into the western hemisphere. There is another, broad and weak lower (4–5km) jet that may form over Africa at the same period of the year and at a similar latitude. It has an important effect on the development of systems that may go on to form tropical cyclones (hurricanes) over the Atlantic. There are various jets that also occur at higher altitudes, such as the Polar Night Jet that forms in the northern winter at a latitude of approximately 60°N and an altitude of about 25km. It is actually the counterpart of the Equatorial Jet Stream, just mentioned, that forms in the summer hemisphere.

Jet streams are not continuous around the Earth. They may show gradual or rapid changes in velocity, sometimes fading away or being regenerated. They may start and stop (abruptly or gradually); split into two (a condition known as split flow); or combine into a single jet. In particular, they usually show distinct variations in latitude. The flow may become northerly or southerly, and may even, on occasion,

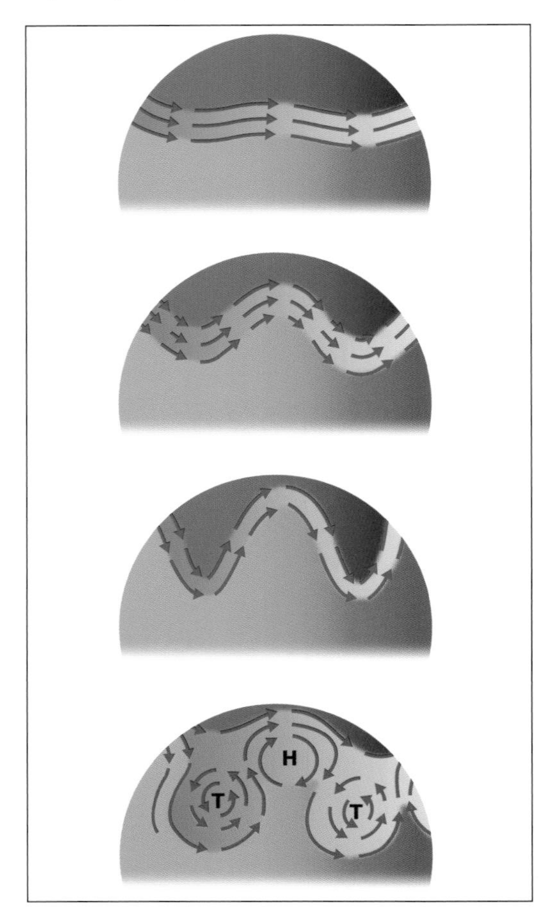

RIGHT The development of meanders and cut-off circulations in the jet stream. *(Ian Moores)*

have short easterly sections. The Polar Jets, in particular, have a major influence on the development, motion, and decay of the weather systems that develop along the Polar Front.

Despite the altitude of the jet streams, they are, to a certain extent, influenced by the terrain beneath them. This particularly applies to the Polar Jet in the northern hemisphere, which is influenced by the Rocky Mountains in western North America, the Ural Mountains and also by the high Tibetan Plateau. A mountain chain that lies across the flow of a jet, even though the latter is in the upper air, not only causes vertical waves, but also creates waves that vary in latitude. It has, for example, been calculated that if the Rocky Mountains did not exist, the Polar Jet would take a more direct course across the Atlantic, with lessened north/south meanders, causing weather systems to acquire less heat from the North Atlantic Current and the North Atlantic Drift, and consequently producing generally colder temperatures in western Europe. Further east, in winter, the course of the jet stream often lies to the south of the Tibetan Plateau and the Himalayas. When its path switches to the northern side of the mountainous massif it helps to initiate the south-western monsoon, although this is primarily driven by intense summer heating of the Thar Desert and neighbouring regions of India.

Convergence and divergence

So far, except when discussing the global circulation cells, we have mainly considered horizontal airflow, *ie* the pattern of surface winds. However, vertical motion also plays an important part in individual weather systems. These vertical motions may arise as a result of basic convection or through convergence or divergence.

When there is heating of the surface, either over the lands and seas of the equatorial zone or over heated land surfaces, convection causes the air to rise, producing low pressure at the surface, and thus giving rise to a warm low, such as that in the equatorial trough. Conversely, air that is cooled over polar regions or in the centres of continents in winter subsides, creating a cold high area at the surface from which the air flows outward.

If air converges at the surface, it obviously cannot just accumulate in that area, so it is forced to ascend. It then diverges at altitude. If no surface heating is involved, this gives rise to a cold low or depression. Conversely, if convergence at altitude forces air to descend, it warms as it does so, and produces high pressure at the surface, in a warm high.

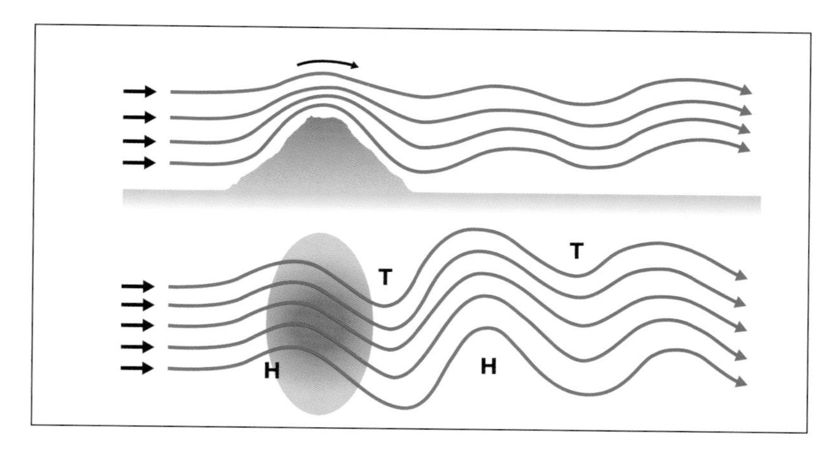

ABOVE Side (top) and plan (bottom) views, showing the effect of a mountain barrier on the motion of the jet stream. *(Ian Moores)*

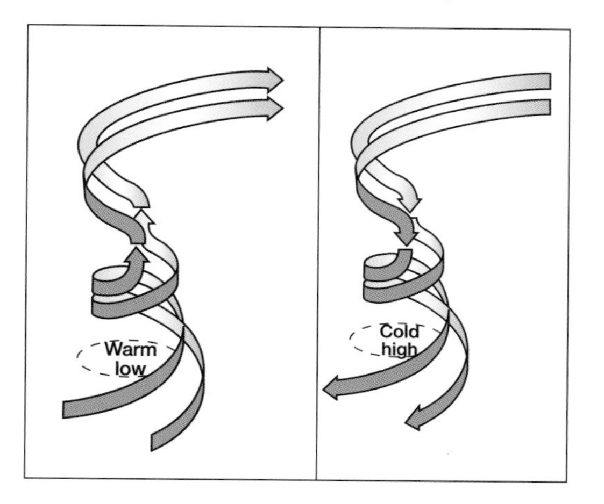

LEFT The overall airflow in a warm low (left) and a cold high (right). *(Ian Moores)*

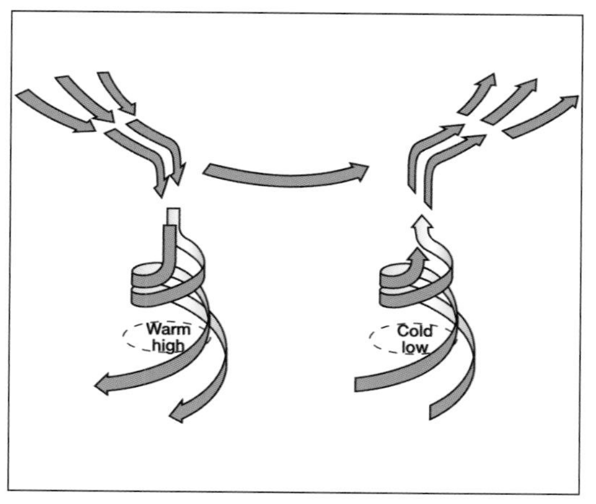

LEFT The overall airflow in a warm high or anticyclone (left) and a cold low or depression (right). The diagram also illustrates the typical convergence at altitude above a warm high, and divergence over a cold low. *(Ian Moores)*

Cross references

Frontal systems p.30
Monsoons p.23
Ozone holes p.80
Tropical cyclones p.124

The long waves in the zonal flow in the upper atmosphere also create regions of convergence and divergence, particularly where jet-stream conditions exist. This convergence or divergence has a strong effect on conditions at the surface and on the growth and decay of both low-pressure and high-pressure regions – that is, on cyclones (depressions) and anticyclones.

The exact effect depends on the altitude of the main zonal flow. At the top of the troposphere, the inversion marking the base of the stratosphere prevents any significant upward motion, so if there is convergence at that level, the air is forced to descend. With convergence in the middle troposphere, some air may ascend and the remainder descend towards the surface. The opposite situation occurs when the air diverges. In the middle troposphere, the potential deficit causes air to be drawn upwards from lower levels. This may lead to the formation or intensification of a low-pressure area (a cyclone or depression) at the surface. Similarly, convergence at upper levels will force air to descend, which will tend to intensify any existing high-pressure area (anticyclone) or, often more significantly, to accelerate the decay of a surface low.

At some altitude (which depends on the exact conditions) the air rising above the centre of a surface low becomes able to diverge: to flow outwards. In some cases, such as major tropical cyclones – which are warm lows – the volume of air that flows from the top of the system is so large that the Coriolis effect comes into play and the outflow is identical to that from a surface anticyclone, and thus clockwise in the northern hemisphere.

In certain cases both mechanisms may play a part, as for example in the Inter-Tropical Convergence Zone, where the trade winds from both hemispheres come together close to the equator. There the motion of the converging air is converted to ascent, assisted by heating from the surface, to give rise to the ascending limb of the Hadley Cell. By contrast, divergence occurs at the surface beneath the subtropical anticyclones that mark the location of the descending limb of the Hadley Cell.

RIGHT Hurricane Katrina on 28 August 2005, just before it devastated New Orleans. *(NASA)*

Air masses

If air is stationary over a particular area of the globe for a moderate amount of time, it tends to assume specific characteristics (in particular consistent temperature and humidity) throughout its volume. It also acquires a specific lapse rate. Such a volume of air is known as an air mass, and the region over which it originates is known as the source region. The principal source regions are the semi-permanent high-pressure zones (the subtropical and polar anticyclones) and the continental anticyclones that develop in winter.

There are four principal types of source region, which are classified in terms of temperature. These (with their standard, single-letter abbreviations) are:

- Arctic and Antarctic (A)
- Polar (P)
- Tropical (T)
- Equatorial (E)

Antarctic air is sometimes designated AA, when it is necessary to differentiate it from arctic air.

There are, in addition, two basic types of air mass: maritime (m) and continental (c), depending on whether the air mass has developed over an area of sea or land. As may be expected, maritime air has a higher humidity than continental air.

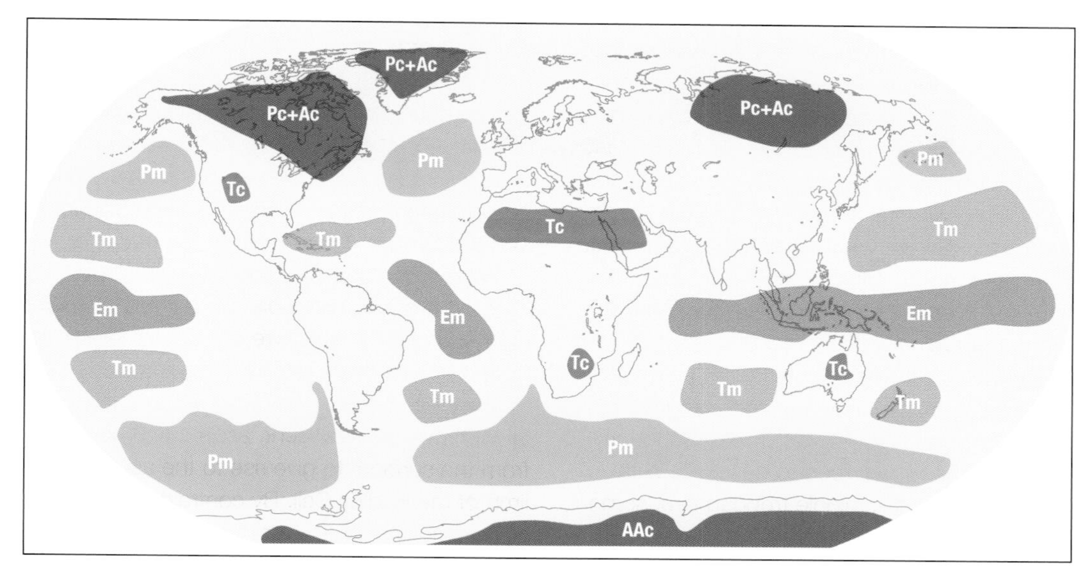

RIGHT Air-mass source regions. *(Dominic Stickland)*

AAc Antarctic continental
Ac Arctic continental
Am Arctic maritime
Em Equatorial maritime
Pc Polar continental
Pm Polar maritime
Tc Tropical continental
Tm Tropical maritime

When these features are combined, we have seven principal categories of air mass (again with the standard abbreviations):

- Arctic continental (Ac) — extremely cold and dry
- Arctic maritime (Am) — extremely cold and humid
- Equatorial maritime (Em) — hot and humid
- Polar continental (Pc) — cold and dry
- Polar maritime (Pm) — cold and humid
- Tropical continental (Tc) — hot and dry
- Tropical maritime (Tm) — warm and humid

Note: In many works, including recent ones, the older notation is used, where the order of the letters and words is reversed, giving, for example, 'maritime polar air (mP)'.

There is no category of equatorial continental air, because there are no land masses in the equatorial region that are able to generate extremely hot, dry air. Antarctic continental air (AAc, extremely cold and dry) is present over Antarctica throughout the year. Arctic continental air (Ac) is, however, found only in winter in the Arctic, when the sea is ice-covered. When the Arctic Ocean is ice-free it produces Arctic maritime air (Am). The other forms of air mass are present throughout the year.

A further important characteristic of different air masses is stability. Polar air is stable, because it is cooled from below by contact with a cold surface. Tropical air, by contrast, is unstable, because it is heated from below. (Stability and instability are treated in more detail later.)

The various source regions are closely related to the semi-permanent pressure features known as centres of action, described earlier. It is here that the air essentially stagnates long enough to acquire its characteristic features. Most centres of action are present (to varying degrees) throughout the year, with the exception of the Canadian and Siberian Highs, which develop in the northern-hemisphere winter.

When an air mass moves away from its source region, it initially retains its characteristic temperature, humidity and lapse rate, but these then become modified progressively with time and distance, depending on the nature of the surface over which it is moving. Air moving over the sea will become more humid, especially in its lowermost layers, through evaporation from the water surface. By contrast, an air mass that has a long track over land, particularly over a large continental area, will remain dry. The temperature of the underlying surface also has an important influence. Cold (arctic or polar) air that moves over a warmer surface will be heated from below. Its lower layers therefore tend to become unstable and subject to convection, whereas in warm (tropical) air that moves over a cold surface, the very lowermost layer cools and becomes stable. There is also a difference in the depth to which change operates: the effect of heating from below on a cold air mass spreads throughout a greater depth than the cooling of a warm air mass,

Cross references

Centres of action p.17
Polar lows p.37
Stability and instability p.50
Subtropical highs p.30
Temperate lows p.30

Cross references

Azores High p.17
Cloud types p.54
Convection p.87
Drizzle p.95
Föhn effect p.108
Lapse rates p.51
Rain shadow p.108
Stability and instability p.50
Supercell storms p.116

which is confined to the layer in immediate contact with the ground. There will, of course, tend to be differences in the effects, depending on the season and whether the air mass has taken a short or long track over land or sea. Tropical continental air approaching the British Isles, for example, will tend to remain very warm or even hot, and relatively dry in summer, whereas in winter it will be cooler and contain more moisture.

Certain regions are also affected by a modified air mass, known as 'returning' air. In the case of the British Isles, where this form of air is often encountered, a polar maritime (Pm) air mass, originally from the area of Greenland and northern Canada, has taken a particularly long track over the Atlantic Ocean, first moving southwards before turning and approaching the British Isles from the south-west. This air, designated rPm, is warmer and somewhat more moist than normal polar maritime air.

The weather at a particular location is primarily determined by the properties of the resulting air mass and the way in which the different air masses interact to give rise to the depressions that are such a significant feature of the weather in temperate latitudes.

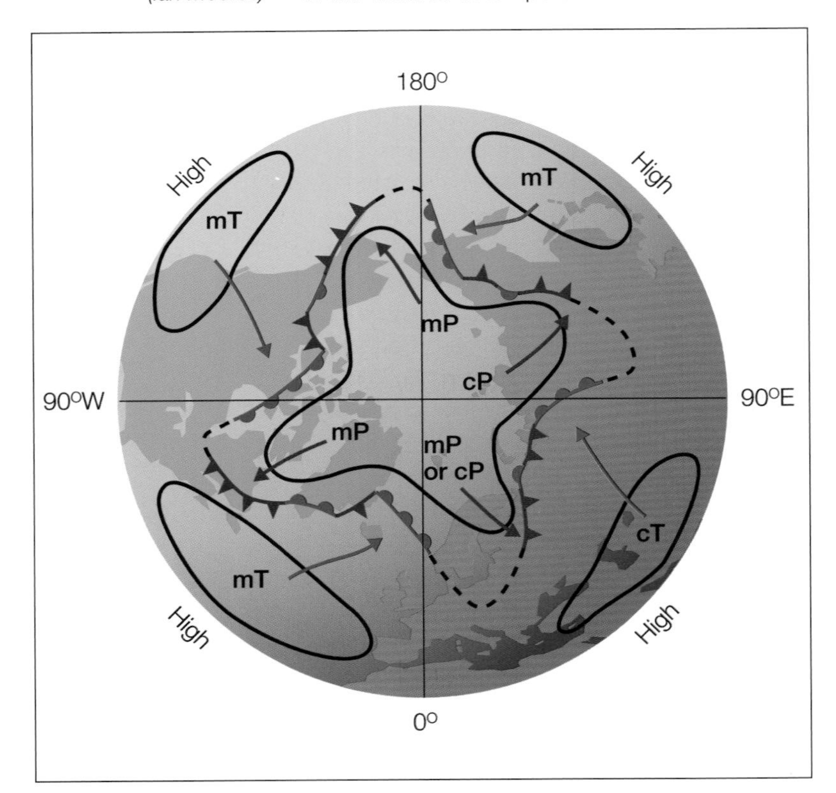

BELOW A typical pattern of the lobes (troughs and ridges), frontal systems and source regions in the northern hemisphere. *(Ian Moores)*

Fronts

The boundary between two different air masses of different temperatures (and usually different humidities) is known as a 'front'. There are three forms: warm, cold and occluded. The first two are determined by the character of the air mass that is advancing, relative to another. The situation regarding the third is more complex and will be described in detail later.

Note that it is the relative temperature difference between the two air masses that is important. In wintertime in the northern hemisphere, there may, for example, be what is sometimes termed an Arctic Front lying between Greenland and Scandinavia, where frigid arctic maritime air borders cold polar maritime air. A more distinct such frontal zone occurs around Japan, where there are greater temperature contrasts than over the northernmost Atlantic. A somewhat similar front – also termed the Arctic Front – may exist over Canada between arctic continental air and polar continental air. In the south, there is a semi-permanent Antarctic Front lying between the frigid Antarctic continental air that flows off the Antarctic continent and the warmer polar maritime air further north.

The two Polar Fronts are particularly important in the generation of weather systems over the temperate regions of the world. This is particularly related to the extent of the meanders that exist dividing cold polar air from warmer subtropical air. When the deviations in latitude are reduced, we have basically westerly winds in what is known as zonal flow. This changes to a meridional flow pattern when lobes of cold air (troughs) push down towards the equator, separated by lobes of warm air (ridges) extending towards the poles. Areas of low pressure (technically known as cyclones, but more commonly called depressions) tend to develop near the regions where the troughs reach the lowest latitudes, and high pressure (anticyclones) where the ridges are closest to the poles. Because of the increased temperature contrast in winter, ridges and troughs are generally more pronounced during that season, with a greater meridional circulation pattern, contrasting with the zonal pattern more common during the summer.

There are typically four to five lobes to the Polar Fronts around the globe. These lobes consist of an alternation of warm and cold fronts, on the eastern sides of ridges and troughs, respectively. The lobes tend to shift gradually eastwards around the Earth, although very occasionally they may move towards the west. There are continual changes as the lobes grow and shrink, and they may occasionally become 'blocked', remaining stationary for considerable periods of time. In extreme cases the frontal boundary may 'pinch off', resulting in isolated pools of cold or warm air, surrounded by a different air mass, giving rise to 'cut-off lows' or 'cut-off highs'. Cut-off lows are depressions that tend to lie at a lower latitude than normal, and the cut-off highs are anticyclones divorced from their subtropical highs and at abnormally high latitudes. The cut-off highs tend to create blocking situations that may be very persistent.

The boundary between air masses is rarely stable and unmoving, but quasi-stationary fronts may persist for some time. Fronts may also develop (in the process known as frontogenesis) or weaken, leading to eventual dissipation (frontolysis). The fronts generally associated with low-pressure areas (depressions) are warm, cold and occluded fronts. All these forms of front, together with upper-level warm and cold fronts (where the boundary between the different air masses does not reach the surface), are shown by distinct, conventional symbols on weather charts and are given (with brief descriptions) in the table. It may be noted that the protrusions (semi-circles on warm fronts and triangles on cold fronts) are always placed on the leading edge of a front, except in the case of a quasi-stationary front, when they alternate. The most common symbols are shown in the diagram, and the full listing is given in the table overleaf.

The development of depressions

The Polar Front is the boundary between the cold polar air and warm subtropical air. The winds on either side of the front may be the polar easterlies and the westerlies, flowing in opposite directions, or both may be westerlies, flowing approximately along the isobars. Such a situation is not stable, with air of different temperatures

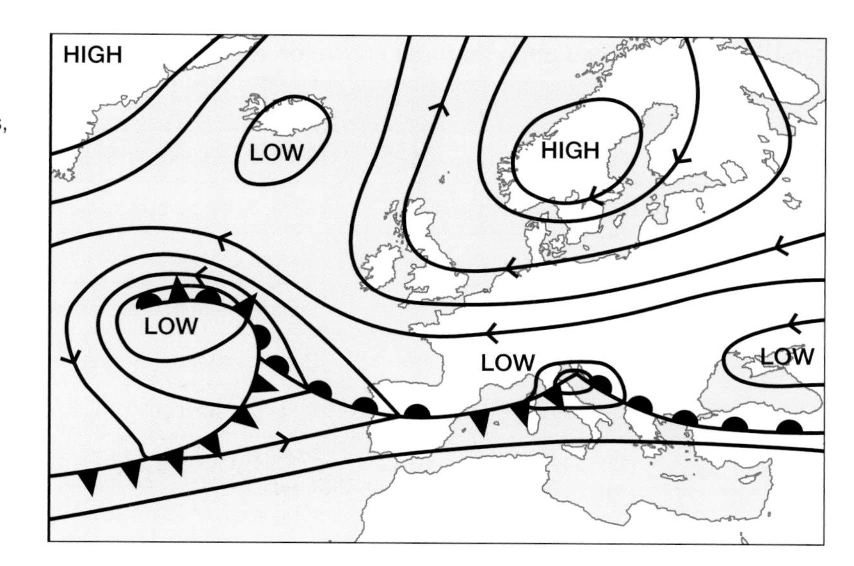

ABOVE A bocking situation over Scandinavia that affects the weather of the British Isles and of neighbouring regions. *(Ian Moores)*

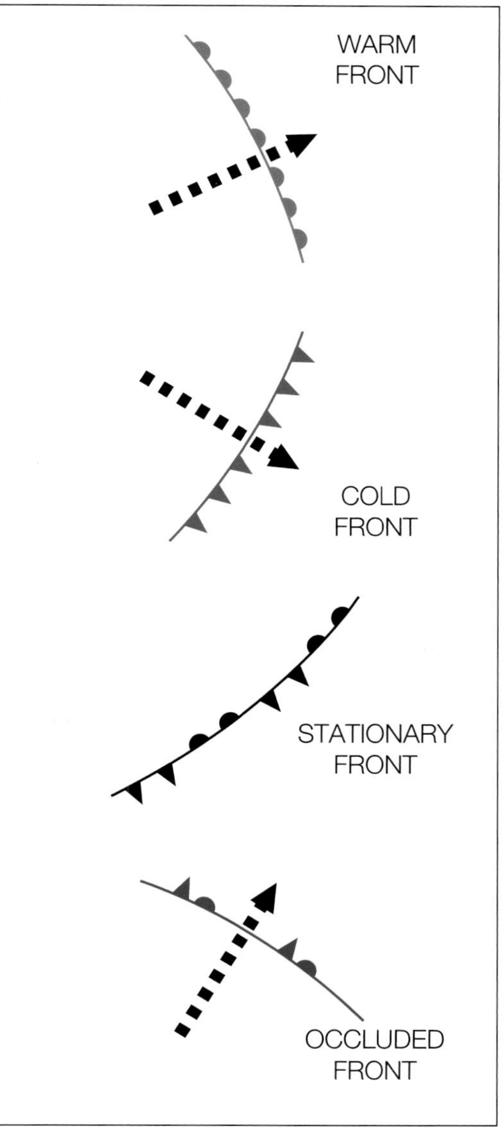

LEFT The internationally adopted, standard method of representing fronts, with arrows showing their typical motion, but which may, depending on circumstances, occur in any direction. *(Ian Moores)*

Symbols for fronts and other features shown on weather charts

Symbol		Description
	Cold front	Surface cold front advancing.
	Cold front frontogenesis	Surface cold front intensifying, normally because of increasing temperature gradient.
	Cold front frontolysis	Surface cold front weakening, generally because of increasing pressure.
	Warm front	Surface warm front advancing.
	Warm front frontogenesis	Surface warm front intensifying, normally because of increasing temperature gradient.
	Warm front frontolysis	Surface warm front weakening, generally because of increasing pressure.
	Occluded front	Occluded front at the surface.
	Cold front above surface	Cold front above surface.
	Warm front above surface	Warm front above surface.
	Quasi-stationary front at the surface	Quasi-stationary front at the surface.
	Quasi-stationary front above the surface	Quasi-stationary front above the surface.
1024	Isobar	Isobar (with associated pressure in hPa).
L × 978	Low-pressure centre	Low-pressure centre (with pressure in hPa).
H × 1024	High-pressure centre	High-pressure centre (with pressure in hPa).
	Trough	Trough.
	Convergence line	Convergence line.
	Ridge axis	Ridge axis.
	Centre of tropical cyclonic circulation (> 64 knots)	Centre of tropical cyclonic circulation with wind speeds of 64 knots or more.
	Centre of tropical cyclonic circulation (< 64 knots)	Centre of tropical cyclonic circulation with wind speeds below 64 knots.
Cross references		*Convergence p.27 Tropical cyclones p.124*

BELOW The normal abrupt change in the direction of the isobars (and the resulting winds) across warm (left) and cold (right) fronts. *(Ian Moores)*

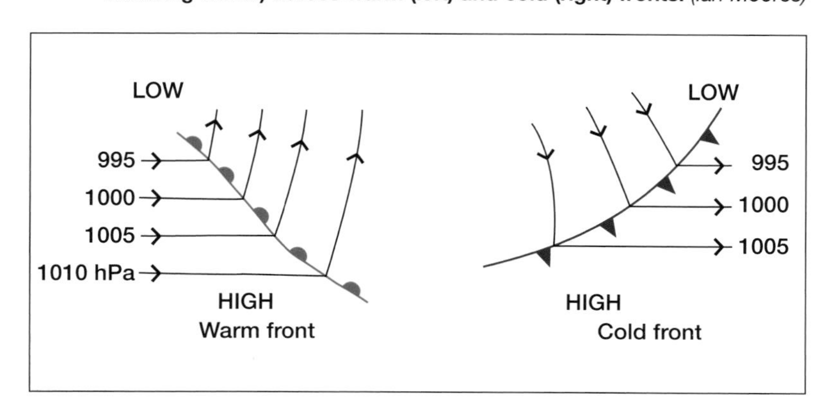

in juxtaposition with one another, and what may initially be a quasi-stationary front that lasts for a few days rapidly develops an irregularity, where warm air bulges into the colder air. This creates a definite wave, known as a frontal wave. This grows until a distinct low-pressure centre forms, with the air flowing across the isobars and with a closed circulation. The system is now known as a 'warm sector depression', with recognisable warm and cold fronts, and distinct changes in the wind direction across these fronts. (In the northern hemisphere the wind veers at the fronts, and in the southern it backs.)

The boundary between different air masses

is never completely sharp. There is a zone, known as the frontal zone, with a vertical depth of 1–2km, where there is slow mixing of the two air masses. In reality, all the fronts have a very shallow slope, which has to be greatly exaggerated in the diagrams. A typical warm front has a shallow slope of between 1:100 and 1:150, and a typical cold front one of between 1:50 and 1:75. With an approaching warm front the tip of the wedge of warm air may therefore reach the tropopause as much as 1,000–1,500km ahead of the surface front. There, the frontal zone may extend over a distance of 100–300km. Similar considerations apply at a cold front where the warm air is being undercut by the cold air, but here the frontal zone is steeper.

At the most active fronts, where there is convergence at the surface, warm air is rising relative to the frontal zone. Such fronts are called 'ana' fronts (from the Greek word for 'up') in what are known as ana warm and ana cold fronts. Provided the air is moist enough, the air will reach the dew point and produce cloud and precipitation. Each front is accompanied by typical types of cloud. These are particularly distinctive at the warm front, giving a recognisable sequence of cloud types as a depression approaches, with high-level cloud first appearing that slowly becomes thicker and closer to the surface, eventually giving rise to prolonged rain. This sequence of clouds and the precipitation associated with the fronts will be described in detail later. Most of the rain associated with an ana warm front will fall, relatively steadily and over a long period, from Nimbostratus cloud.

At an ana cold front, the sequence of clouds will normally be the reverse of that at a warm front. Because the front is steeper, there will be a relatively sharp cut-off behind it as the cold air arrives, together with a marked veer in the wind. In many regions, including over the British Isles, the air in the warm sector is generally stable, so the main rain-bearing cloud at this front is Nimbostratus. Elsewhere, a 'classical' ana cold front may occur with unstable air giving rise to deep convective clouds, heavy rain of shorter duration, and thunderstorms. In either case, the cold air behind the front may be unstable and accompanied by Cumulonimbus cloud and showers, particularly if the cold air is heated from beneath as it passes over relatively warm seas. However, over continental areas in winter, the cold air behind the front may receive little heating from the surface and so does not develop deep convective cloud. This is often the case over the central United States in winter, for example, when cold air sweeps south from the Canadian Arctic.

However, as we have seen, there may be convergence in the middle troposphere and descending air, causing the warm air to be sinking relative to the frontal zone. This leads to subdued fronts, with a different, but in both cases a somewhat similar, sequence of clouds, marked in particular by thick Stratocumulus. Such fronts are known as 'kata' fronts (from the Greek word for 'down'). At a kata warm front, the sequence of clouds, beginning with high Cirrus and descending to Nimbostratus, is missing, replaced by lower cloud gradually thickening to Stratocumulus, giving only light rain or drizzle. The Stratocumulus may thin slightly behind the front, and merely thicken

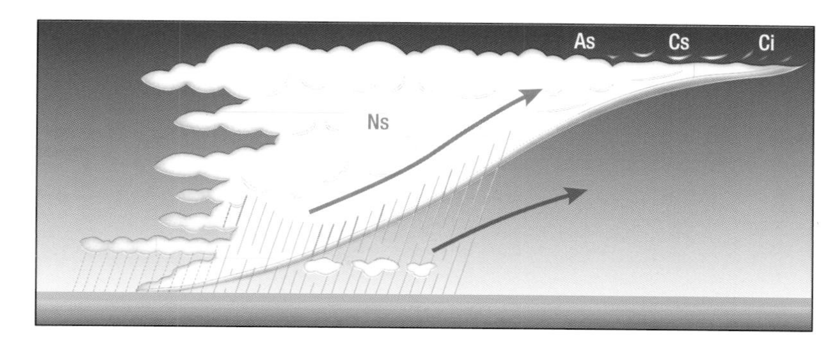

ABOVE A 'classic' ana warm front at which the warm air is rising over the whole frontal zone. *(Dominic Stickland)*

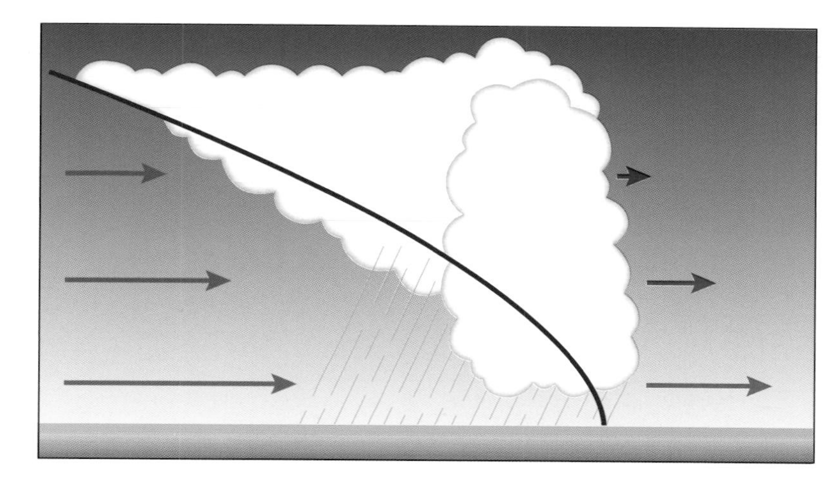

ABOVE An ana cold front where the cold air is undercutting the warm air and forcing it to rise. *(Dominic Stickland)*

Cross references

Cirrus p.60
Cloud classification p.54
Cumulonimbus p.73
Cumulus p.70
Nimbostratus p.67
Showers p.111
Stability and instability p.50

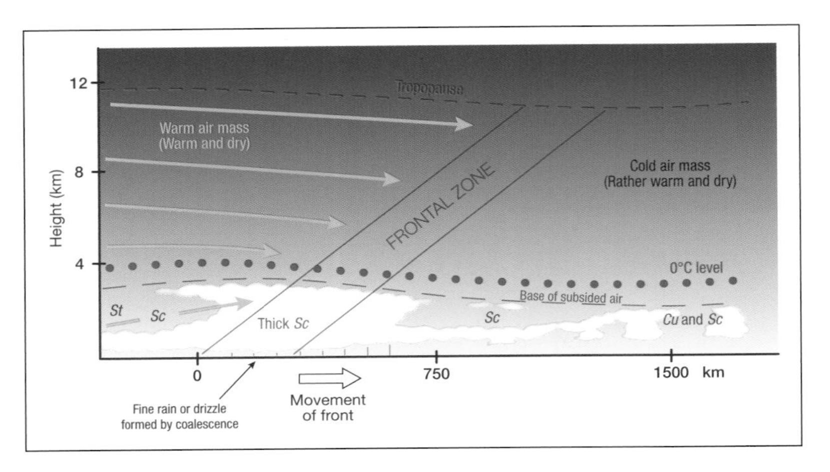

LEFT A kata warm front, where the air in the warm sector is subsiding and tending to suppress the formation of deep clouds. *(Dominic Stickland)*

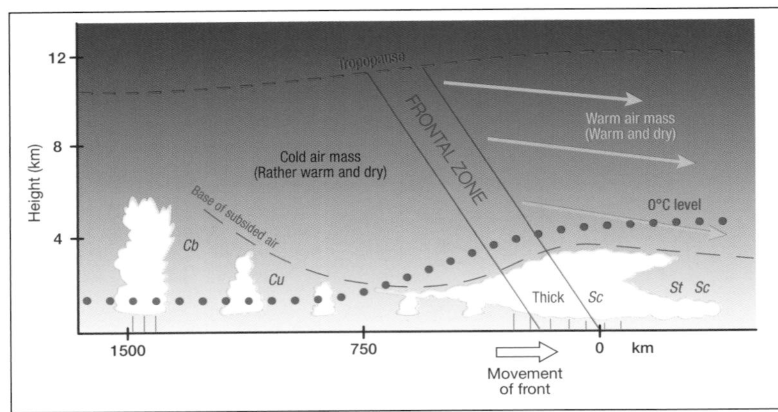

ABOVE A kata cold front, where the subsiding air ahead of the front tends to produce thick Stratocumulus cover in the warm sector. *(Dominic Stickland)*

again at the kata cold front, where convective cloud is completely absent. Again, any rainfall is very light. In the cold air behind the cold front, Cumulus and Cumulonimbus cloud may form in unstable air.

Frontal systems are rarely clear-cut and many show a complex mixture of features, with sections where the warm air is rising and others where it is sinking. The characteristic features may also vary over time, leading to extremely varied frontal systems, no two of which are ever truly alike.

The advancing cold air tends to move faster than the warm air, so the cold front in a depression overtakes the warm front, gradually narrowing the warm sector of the depression. When it reaches the warm front, it starts to lift it away from the surface, and comes in contact with the cooler air that was originally ahead of the warm front. This produces what is known as an occluded front, where a pool of warm air has been lifted above the surface. An isobaric chart exhibits three different forms of front, and the point where they meet is known as the 'triple point'.

An occluded front may be one of two types, depending on the relative temperatures of the two cold air masses. In the most common case, the air behind the front is the colder, so it tends to undercut the other cool air in what is a cold occlusion. Less frequently, when that air is warmer, it tends to override the air ahead of the front. Once it has reached the occluded state, the depression decays and the closed circulation breaks down.

Once a closed circulation has become established, the whole system moves eastwards, partially under the influence of the jet stream, which (if present) will lie nearby and above it. An isobaric chart for a higher altitude (say for 300hPA) shows how the jet stream itself

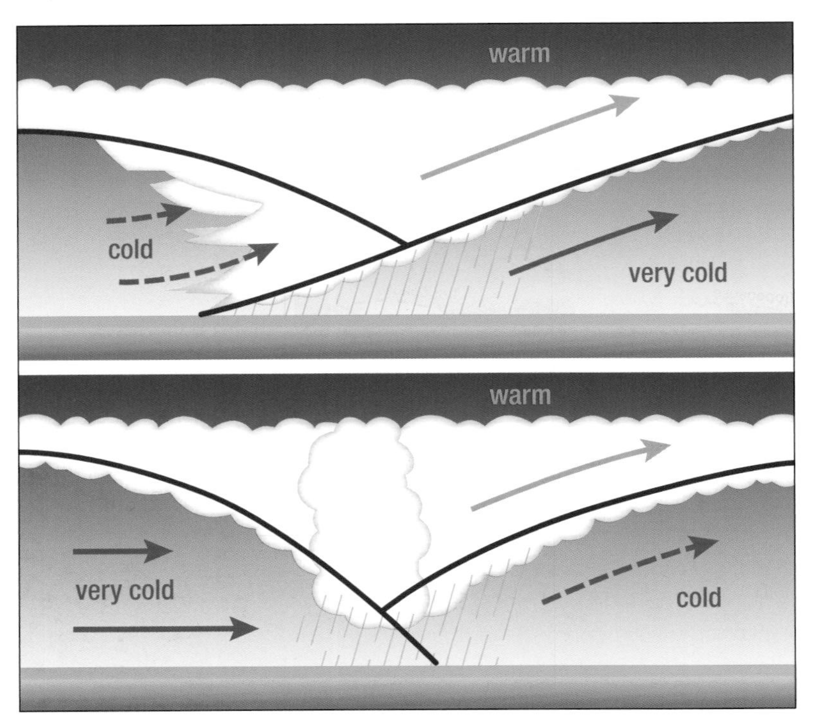

LEFT A warm occlusion (top) where the coldest air is ahead of the front, and a cold occlusion (below) where the coldest air follows the front. *(Dominic Stickland)*

develops a wave above the depression and will be flowing generally towards the north-east behind the cold front and towards the south-east ahead of the warm front.

As one depression matures and deepens and moves east, the trailing cold front behind it pushes further towards the equator (south in the northern hemisphere). It is prone to develop secondary waves, and these may travel swiftly along the front. Such cold-front waves may never develop into a full warm-sector depression. They do, however, bring their own area of precipitation, and they tend to slow down the progression of the cold front towards lower latitudes. Many, however, do become independent depressions, which tend to evolve separately to the original low that lies further to the east and north. There may, therefore, be a succession of waves and depressions, in different stages of development, following one another across the globe. Generally, each subsequent depression will form at a lower latitude than the preceding one, until eventually the sequence breaks off, and the front retreats to a higher latitude. As the leading depression starts to decay a nearby following secondary depression may begin to overtake it on the equatorial side. In some cases the two systems start to circle one another, with the decaying

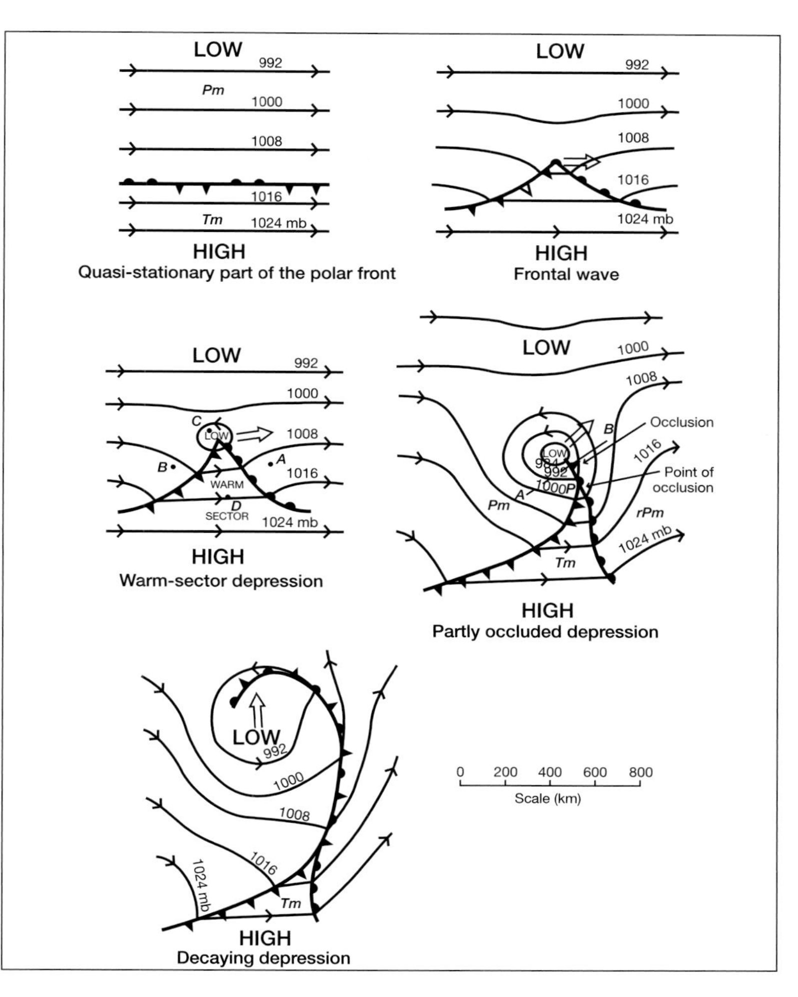

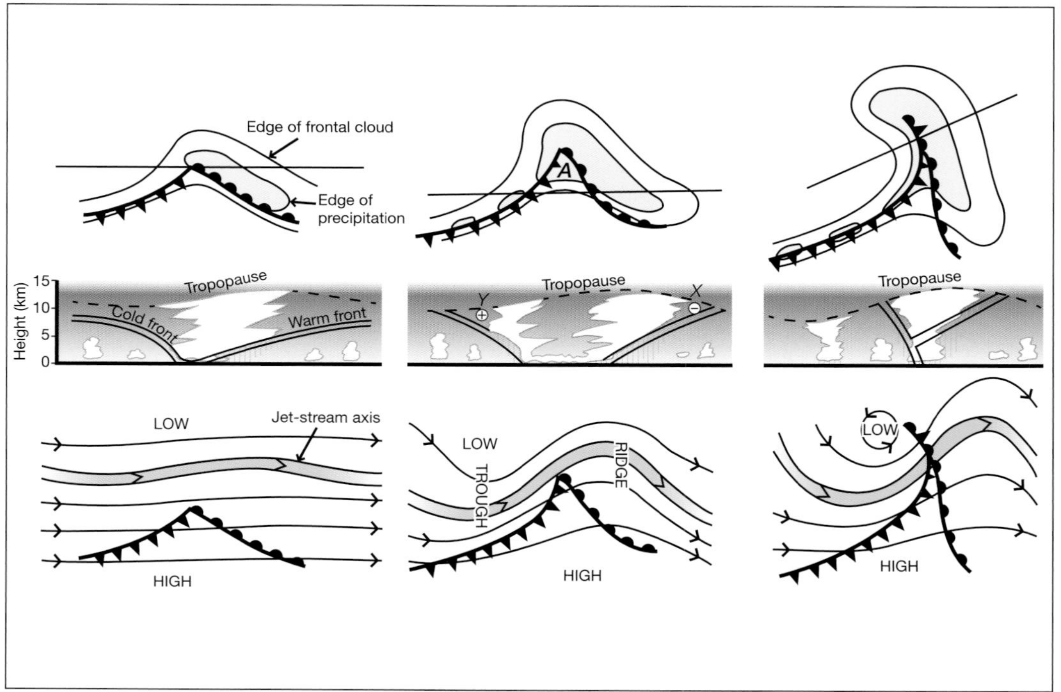

ABOVE **The various stages in the development of a frontal wave into a closed-circulation and finally into an occluding depression.** *(Ian Moores)*

LEFT **The changes in size and location of the rainbelt (pale blue) as a depression develops (top). The development of a trough and ridge in the jet-stream flow as a depression gradually evolves (bottom).** *(Ian Moores)*

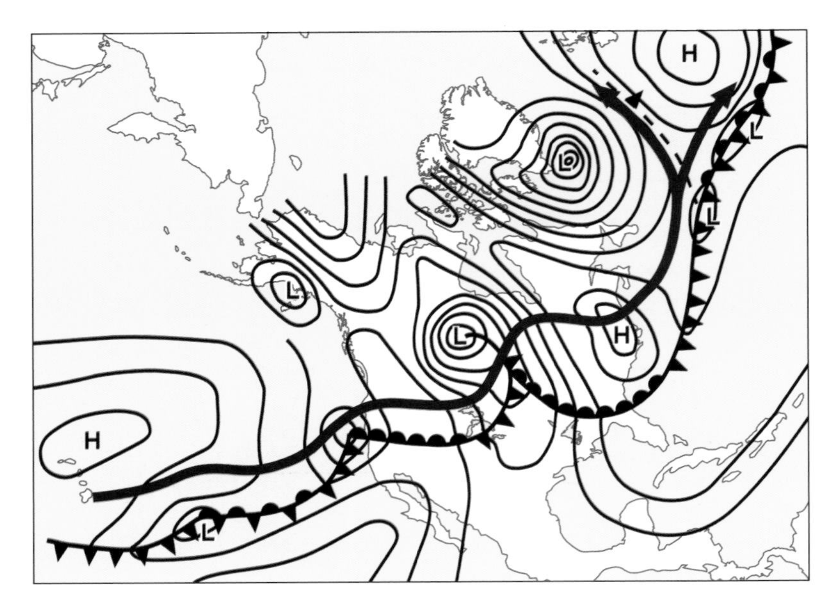

ABOVE A family of depressions, where the easternmost has almost completely decayed, but where one depression is fully formed and is followed by two more that are developing from waves on the Polar Front. The heavy black line shows a typical path of the associated jet stream. *(Ian Moores)*

system moving westwards or even, on rare occasions, westwards and then slightly towards the equator as it dissipates.

Where the jet stream is fast-moving, it tends towards zonal flow, with weak troughs and ridges. A whole succession of depressions may form, one after the other. They decay quickly as they are transported rapidly towards the east. Where the jet-stream winds are slow, there is a much greater tendency for meanders to form and for the troughs and ridges to be accentuated, extending towards the equator and poles, respectively. Large, long-lasting and severe depressions tend to develop. Both they and high-pressure areas are much slower-moving and may linger over a region for a considerable time. In extreme cases, areas of high or low pressure may become cut off from the general flow, affecting the regions below them for extended periods of time.

On fairly rare occasions, waves may develop on the warm front ahead of a major depression, particularly when the cold air mass ahead of the warm front is particularly cold, as occurs occasionally in winter with arctic maritime or polar continental air. Such warm-front waves migrate along the front away from the main depression and are known as break-away lows. When such waves move out of the circulation of the main depression, they start to draw the extremely cold air into their own circulation, and may then rapidly grow into a separate depression.

Airflow and precipitation in depressions

The diagrams showing a section across a frontal system give the impression that the air behind the warm front is flowing directly towards it. However, it is obviously impossible for the air to accumulate indefinitely and, in fact, the airflow alters. The component of the flow that is perpendicular to the front decreases and, correspondingly, the component parallel to the front increases. Motion parallel to the front is actually greater at all levels, and towards the top of the troposphere it may amount to as much as ten times the component towards the front. So the airflow at high levels is parallel to the front. This is often seen from the structure and motion of the highest clouds.

The principal airflow producing cloud and precipitation in a depression is known as the warm conveyor belt. It starts fairly low in the warm sector, gradually ascending from an altitude of about 1km, runs approximately parallel to the cold front, and eventually turns (to the right in the northern hemisphere) and passes over the warm front at an altitude of between 5 and 6km, where it merges with another stream of drier, colder air that originates in the middle troposphere behind the cold front.

LEFT The complex system of 'conveyor belts' found in a depression system. The warm conveyor belt is above the warm front. The pale blue area shows the area of heaviest rainfall. *(Ian Moores)*

It is the warm conveyor belt that transports most of the heat and humid air within the depression system. They have been found to be associated with what are termed 'atmospheric rivers' of intense moisture-laden air originating in the tropics, and will be described shortly.

There is also a cold conveyor belt of air ahead of the warm front that approaches the front before turning (again to the right) to flow approximately parallel to the front, beneath the warm conveyor belt, before rising into the occluded front.

The precipitation at the warm front is spread over a wide area, often as much as 200–300km ahead of the surface front, within which there are bands of lighter and heavier rain. By contrast, precipitation at the cold front is confined to a narrower area with an initial band of lighter rain, followed by a belt of heavier precipitation some 50km wide immediately ahead of the front. In both cases there are individual cells with the heaviest precipitation within the bands of heavier rain.

Thermal lows

There are two forms of non-frontal low-pressure regions that may have a significant effect on local weather. The first is known as a thermal low or thermal depression, and is likely to develop in summer when daytime heating over land is at a maximum. It may lead to the formation of a low-pressure centre with closed isobars and circulation. On many occasions the heating is insufficient to create a closed circulation, and it merely tends to distort the existing isobars in what is termed a thermal trough. Such thermal lows and troughs tend to dissipate at night, when surface heating disappears. They may, however, depending on the air mass within which they occur, create sufficient instability to initiate showers or even thunderstorms during the day. In the tropics in summer, the rise in pressure during the night may not be sufficient to offset the decline in pressure during the day, leading to a semi-permanent thermal low. The prime example of this is the thermal low that occurs over the Thar Desert and surrounding area of India, which strongly influences the development of the summer monsoon.

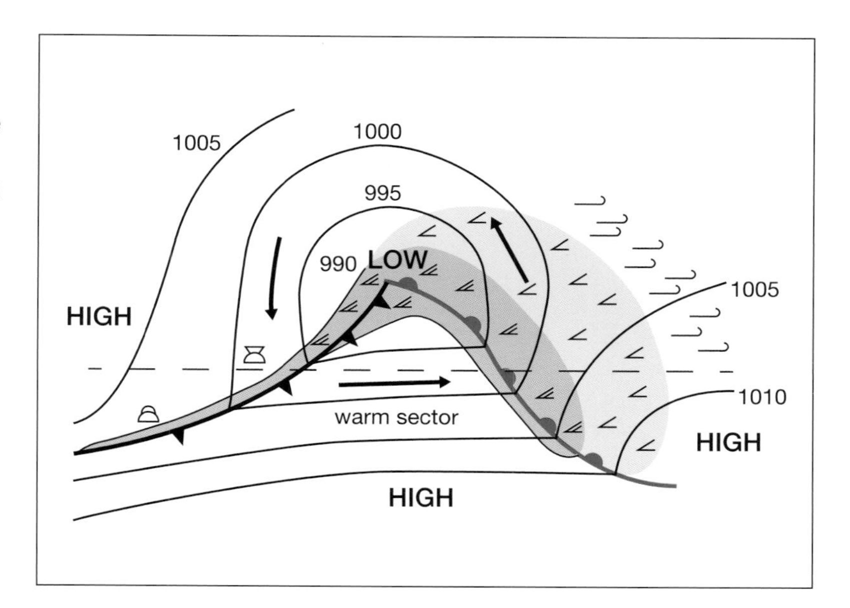

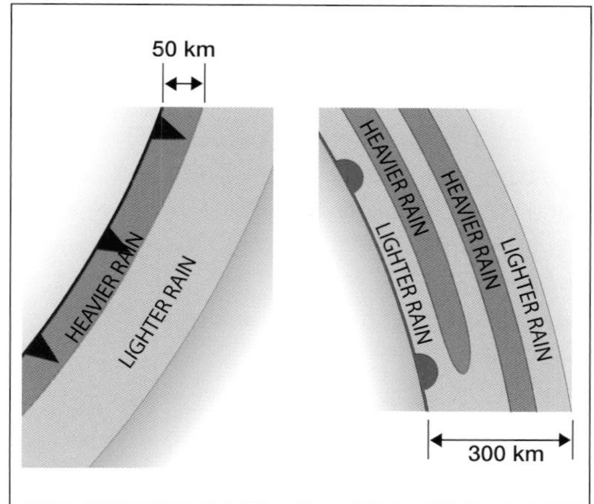

ABOVE The overall pattern of rain in a depression, before any occluded front has formed. *(Ian Moores)*

LEFT The broad belt of rain ahead of a warm front normally contains bands of heavier rain. At a cold front, by contrast, although it is preceded by light rain, the heaviest precipitation is close to the front itself. *(Ian Moores)*

Polar depressions

A thermal low may also be created by relatively intense heating over the sea, especially when polar air flows over open water. This results is rapid heating of the lower layers by the warm sea, and the greatest effect occurs in winter when there is the greatest contrast between the air and water temperatures. In addition, and in contrast to the thermal lows that arise over land, heating is continuous throughout the day and night, so such polar depressions (also known as polar lows) may become very intense. As with thermal lows over land, weaker heating may produce a polar trough, rather than a closed circulation.

The most common location for the formation of a polar low (or lows) in the northern hemisphere is in northerly airflow on the western

Cross references

Monsoons p.23
Showers p.111
Stability and instability p.50
Thunderstorms p.114

ABOVE A well-developed polar low over the North Atlantic, south of Iceland. *(NASA)*

side of large occluding depressions. The polar maritime air that is streaming south rapidly becomes very unstable, so the troughs or lows are the site of intense convection, with heavy showers that may sometimes merge to give prolonged periods of rain or snow.

Atmospheric rivers

During the late 1990s, with the introduction of microwave sensing on meteorological satellites, it was discovered that large amounts of moisture-laden air are transported in what have come to be called atmospheric rivers, narrow streams in the middle troposphere, perhaps 300–400km across, that may extend for thousands of kilometres across the globe. These intense bands have been found to carry vast quantities of water at any one time: as much as 20% of the total water transport from the tropics to higher latitudes. These rivers of moisture had been undetectable with earlier satellites' infrared sensors. Although their exact causes have yet to be determined, they appear to arise when deep low-pressure regions draw the moisture-laden air into a narrow stream ahead of the cold front, which then later becomes the previously recognised warm conveyor belt, described earlier.

At any one time there seem to be five or six of these atmospheric rivers snaking their way from the equatorial region towards middle latitudes. The amount of moisture that each one carries has been likened to the flow of the Amazon, or about ten times that of the Mississippi. Most of these rivers lose their precipitation over the oceans, and normally even when they reach land the variations in location caused by various existing weather systems tend to distribute the rainfall or snowfall over a wide area. Occasionally, however, stationary weather systems, such as an anticyclone, force the rivers to flow over one specific area. If such a river encounters a mountain range, for example, it may deposit exceptionally large quantities of precipitation in one small region. Such occasions are now known to be responsible for major episodes of flooding in California (where the airstreams are forced to rise over the Sierra Nevada, which parallels the coast, some distance inland). Similar situations have also led to devastating floods in Britain in Cumbria in 2009 and in Cornwall in 2012, although in both cases, because the atmospheric river had crossed the relatively cool North Atlantic, the quantity of moisture transported and the subsequent rain was less than was experienced in California with the atmospheric rivers that arose over the relatively warm Pacific. In 2010 an atmospheric river approached the eastern seaboard of the United States, where it encountered a powerful squall line of thunderstorms, over which it was forced to rise, causing it to deposit most of its moisture over Tennessee, with 300–500mm of rain over and around Nashville. There was widespread flooding, and 11 deaths in Nashville itself, where the Cumberland River overflowed its banks.

Now that these atmospheric rivers are known to exist, they may be identified in satellite images and incorporated into forecast models. With the installation of specific atmospheric-river observatories and instrumentation on the West Coast of America, forecasters are able to predict the strength and location of rainfall a few days in advance. Such methods will undoubtedly be extended to other parts of the world, and in the meantime intense research is being carried out to try to understand the mechanisms involved in their formation.

Chapter 4

Low- and high-pressure systems

An approaching depression

The first sign of the approach from the west of an active depression (where the air is rising at both warm and cold fronts) may be the appearance of jet-stream Cirrus high overhead. At this very early stage, the jet stream is flowing from a generally north-westerly direction and very approximately parallel to the distant warm front. More commonly, however, the first indications of a depression are wisps of Cirrus that gradually increase from one direction and eventually merge into a sheet of Cirrostratus. Similarly, aircraft contrails become persistent and may spread to cover a large area of the sky. Both of these signs are indications that the air high above is becoming more humid as the front approaches. Once a more or less continuous sheet of Cirrostratus has arisen, various halo phenomena are often visible, although they may persist for only a short period, depending on how fast the Cirrostratus itself thickens.

There are other indications that a depression is approaching. One may be described as the 'crossed winds' rule. If you are facing the front (in the northern hemisphere), any low-level wind (and the lowest clouds) tend to move from left to right. Medium-level clouds move directly towards you, whereas high Cirrus (and especially the jet stream and its accompanying Cirrus) may move from right to left. The winds veer (*ie* change in a clockwise direction) with increasing height. This is an indication that the depression is directly approaching the observer and that, in all probability, the warm front, the warm sector, the cold front and the subsequent cold air mass will pass, in sequence, over the observer.

Conversely, if the winds at the different levels tend to back (*ie* change anticlockwise), the weather is likely to improve. Naturally, in the southern hemisphere the change in wind direction is in the opposite sense. In the south, winds that back with increasing altitude are thus a sign of an approaching depression.

Another factor is the rate at which the change from initial Cirrus to thick Cirrostratus takes place. If the alteration takes just a couple of hours, the warm front and its accompanying rain is likely to arrive within a few hours. If the change is slow, taking the best part of a day, the bad weather may not arrive for one or two days.

An approaching warm front

As we have seen earlier, the slope of a warm front is between 1:100 and 1:150 (between 1% and 0.67%), and the typical height of the base of Cirrus clouds is around 20,000ft (about 6km). So when we see the first signs of thickening Cirrus high overhead, the surface front is likely to be some 600–900km distant.

BELOW **High-level Cirrus clouds, showing considerable wind shear aloft, ahead of a warm front approaching from the west (left).** *(Author)*

ABOVE Nimbostratus clouds producing heavy rain at a warm front. *(Author)*

Normally, of course, the front will be to the south-west or west of the observer's position. Depressions tend to advance at approximately the same rate as the speed of the geostrophic wind in the warm sector. In general the average speed is about 50kph, so rain may be expected in about 9–10 hours, and the surface warm front will arrive in about 12–18 hours.

In the relatively cool air ahead of the warm front there will often be Cumulus clouds that gradually evolve into flattened Cumulus humilis as the warm air begins to spread overhead. The sequence encountered in the actual frontal clouds is very distinctive: high Cirrus, Cirrostratus (usually with halo effects), Altostratus, and Nimbostratus. This succession of clouds thickening and lowering is very common, although well away from the depression centre; some Cirrocumulus and Altocumulus may also appear for a short period, slightly interrupting the normal sequence.

Naturally, as these clouds gradually thicken the sky becomes darker. Early Altostratus may be thin enough for the Sun to appear as if seen through ground glass, but soon even this will disappear, and objects on the ground will no longer cast distinct shadows. Although thick Altostratus may produce some precipitation, this rarely reaches the ground. The change from thick Altostratus to Nimbostratus is indicated by the arrival of the first of the true rain bands. The base of Nimbostratus is normally uneven because of the precipitation, which may cool the underlying air sufficiently for it to reach the dew point and give rise to ragged wisps of pannus between the Nimbostratus and the ground.

The duration of the main belt of rain is, of course, dependent on the speed at which the depression is moving, but is generally about 3–4 hours. Rather than being a uniform, continuous belt of rain, there are normally bands of heavier precipitation that run approximately parallel to the surface warm front. Occasionally there may be pockets of instability that produce convective cloud within the overall stratiform clouds, resulting in intense showers and small areas of even heavier rain.

If the air mass ahead of the warm front is very cold, the precipitation may be in the form of freezing rain or snow, before more general rain. When the air is particularly cold the whole precipitation may, of course, be as snow.

As the sequence of clouds progresses, the atmospheric pressure gradually declines because the centre of the depression is approaching the observer. As the warm front arrives, the pressure steadies at its lowest point. The surface wind, which often has a tendency to back slightly ahead of the front, but is at the same time strengthened more or less continuously and may have become gusty, veers sharply with the passage of the front, changing from generally southerly to south-westerly or even westerly. As the front passes, the rain eases or ceases entirely, and the cloud cover thins and may break up. (Naturally, in the southern hemisphere the wind veers slightly ahead of the front, and then backs sharply at the warm front itself.)

Because of the slope of the warm front, the frontal zone (which is about 1km thick) actually extends for as much as 150km at the surface, so it may take a considerable time to pass over the observer. Eventually, however, the warm sector arrives and there is normally a distinct change in the cloud cover and accompanying weather. However, the exact changes will depend on the observer's location relative to the depression's centre or to the triple point (if one has developed). Close to the centre, the rain may persist and merge with the rainfall at the cold front, only ceasing once that front has passed. Similarly, even if the rain does not continue there may be more or less continuous

Cross references

Cloud types p.54
Geostrophic wind p.21
Pannus p.74
Showers p.111
Veering and backing p.16
Warm sector p.32–35

RIGHT Clouds breaking up and thinning after the passage of a warm front. The Nimbostratus has lifted into thin Altostratus, leaving some remnants of pannus in its wake. *(Author)*

cloud cover across the warm sector, and it may be in the form of Stratus or Stratocumulus.

Further away from the centre the clouds may be well broken, with fairly clear skies and warm, pleasant sunshine. In particular, during summer and over the land, there may be sufficient heating for large Cumulus or Cumulonimbus to develop, and these may produce showers or even develop into thunderstorms.

With the arrival of the cold front, the weather changes once again. Such a front is steeper – between 1:50 and 1:75 (2% and 1.3%) – and faster-moving than a warm front. It is generally the site of marked convection, with resulting heavy rain and possibly thundery activity. There is often a band of rain ahead of the front itself. Generally the approach of the front is masked by cloud in the warm sector, but occasionally – and especially well away from the depression's centre – the cold front appears as a distinct line of mixed clouds approaching the observer.

As the cold front passes the wind veers yet again, from south-west or west to west or north-west. The temperature drops – sometimes very dramatically – and the

ABOVE An approaching cold front on a comparitively rare occasion when there was little cloud in the warm sector, allowing a clear view of the front and its associated convective cloud. *(Author)*

LEFT The rear of a cold front, passing away to the east (left), showing the large amount of high Cirrus that may sometimes trail behind such fronts. *(Author)*

ABOVE A satellite image (obtained with amateur equipment) of two depressions. The larger has a poorly defined warm front and warm sector, but shows a distinct 'clear slot' behind the cold front, which is missing from the smaller and weaker, earlier depression at top right. Both show extensive convective cloud in the cold polar maritime air. *(Author)*

pressure starts to rise. (This time the frontal zone is narrower, so the changes take place more quickly.) Immediately behind the cold front there is often a 'clear slot', free of any cloud, and this may often be seen clearly in satellite images. The cold air mass brings good visibility, and is often very unstable, especially if it has approached over the open sea. The instability gives rise to numerous Cumulus and Cumulonimbus clouds, with the latter frequently developing into thunderstorms. However, if the cold air has approached over a continental land mass, such as over the interior of North America, the air does not normally undergo considerable surface heating, so there is less convective activity and fewer resulting showers.

An occluded front

If a triple point has already developed and it is the occluded front that passes over the observer, there are, obviously, differences in the sequence and nature of the events that occur. Because the cold front has caught up with the warm front, and lifted a pool of warm air away from the surface, stratiform clouds typical of a warm front are, in effect, followed directly by the convective clouds of a cold front. In most cases it is a cold occlusion, where the air behind the front is the colder of the two air masses. In such a case the wind tends to veer very sharply (in the northern hemisphere), sometimes by as much as 45°, and there is a sudden drop in temperature as the front passes overhead. Despite being apparently simpler, occluded fronts may still produce large quantities of precipitation, and because of the way in which they may trail behind, or curl round their parent depression, observers may find one passing overhead for hours or days, usually resulting in persistent cloudy skies and a large amount of precipitation. Or else the occluded front may pass away, followed by a relatively clear interval, perhaps with some convective cloud, only for the front (usually weaker) to return as the depression centre finally moves away to the east.

In general, warm occlusions (where the colder air is in advance of the front) are less distinct and less active than cold occlusions, and do not have such a long lifetime.

North of a depression centre

If (in the northern hemisphere) the centre of the depression passes to the south of the observer, there will, obviously, be a different sequence of events. Initially there may be high Cirrus that gradually thickens into Cirrostratus, which may lead observers to assume that a warm front is approaching. Noting the direction of the surface and upper winds, however, will reveal the true situation. The two winds, although approximately parallel with one another, will be blowing in opposite directions. Any cloud does not continue to increase, but

ABOVE A well-developed depression with a long, spiralling occluded front over the Bay of Alaska in the northern Pacific. The image clearly shows the large area of convective cloud following the cold front. *(Dundee Satellite Receiving Station)*

instead tends to disperse, as the pressure drops slightly. The surface wind backs from (say) south-east to east and finally swings round to north-east or north, and the pressure gradually rises as the centre of the depression passes to the south.

South of a depression centre

Well to the south of a depression's centre, although Cirrus may increase into Cirrostratus and the latter into Altocumulus, the cloud does not, in general, completely cover the sky. Any pressure and changes of wind direction are minor and slow, and the pressure soon begins to rise. Frequently there is a ridge of high pressure behind the warm front that prevents any readily identifiable cold front from passing over the observer.

Isolated warm and cold fronts

Both warm and cold fronts may occur on their own, not associated with any particular low-pressure system. These tend to occur more frequently over continental areas (such as the interior of North America or Eurasia) than over maritime regions. They display the classic features of warm or cold fronts, although warm fronts generally display more convective activity. Isolated cold fronts may be very abrupt and typically display a

ABOVE The clearing sky behind a weak depression (in this case with some higher Altocumulus clouds), illuminated by early-morning sunlight. *(Author)*

sharp drop in pressure ahead of the front itself, followed by a sudden rise as it passes overhead.

Anabatic and katabatic fronts

The sequence just described applies to what may be termed 'classic' fronts – anabatic fronts – where the air is rising at both the warm and cold fronts. Sometimes, as we have seen earlier, the warm air mass is subsiding and producing katabatic fronts. Such a low-pressure system is weaker and less distinct. The changes that occur are generally simpler. Because the air is subsiding, it tends to suppress convection, and it does not produce the sequence of high-level Cirrus, Cirrostratus and Altostratus clouds that are such a useful sign of an approaching warm front. Instead, any Cumulus ahead of the system tends to thicken into Stratocumulus, which may become very thick but does not produce prolonged or heavy rain. Any precipitation is usually very light, and is often just in the form of drizzle.

The changes in pressure and wind direction are similar to those in a 'classic' system but are normally less extreme and take place more slowly. Behind the warm front the Stratocumulus usually thins and allows patches of blue sky to be seen, before turning into low Stratus. At the cold front the Stratus and Stratocumulus thicken again and are accompanied by light rain or drizzle. The decrease in temperature and veer of the wind at the front are slower and less extreme. Behind the front the stratiform cloud clears and is replaced by Cumulus and Cumulonimbus in the cold air mass.

Although these low-pressure systems may seem dull and relatively uninteresting, they are very common in many parts of the world, especially during the winter months. Without further information, such as that provided by analysis or forecast charts, it is difficult to estimate the amount of precipitation that any system is likely to produce, nor how fast it will move. One sign, however, is that if the sky becomes dull and dark this is an indication that the cloud cover has thickened, and is probably dense enough for the cloud droplets to grow into larger raindrops. In addition, the speed at which any high or medium clouds advance provides a clue as to whether rain is likely to arrive within a day. If the changes are very slow, the rain belt may never reach you.

High-pressure areas

As mentioned earlier, high-pressure regions are of two types: the cold highs and the warm highs. Both, of course, exhibit divergence at the surface, and are a source of cold or warm winds respectively. The cold highs are shallow anticyclones that form primarily during the winter and particularly in the polar regions or over continental interiors. The air within them may be very cold and the sky tends to be completely clear. On other occasions there may be an inversion at some height above the surface, which limits the upward growth of any Cumulus clouds, which then spread out into Stratocumulus. In other areas of the world the high pressure tends to be in the form of a ridge, rather than an area with a closed circulation. In either case, the air temperature usually drops extremely rapidly at night.

In the other form of anticyclone, the warm high, because air is descending throughout the depth of the troposphere this normally suppresses cloud formation. However, there may be isolated Cumulus clouds, or some broken Stratocumulus. If the air is an extension of a warm, humid, tropical maritime air mass, then large areas of Stratus or fog may form, especially when the temperature falls at night. Such cloud and fog may be extremely persistent, especially during the winter and if the winds are light. In the summer, daytime heating is usually sufficient to disperse the low cloud or fog. Occasionally the overcast skies with thick Stratus and Stratocumulus persist for days or even weeks. Such depressing conditions are known as 'anticyclonic gloom'.

The deep column of descending air found in warm highs acts as a barrier to both the normal westerlies and to the eastward movement of depression systems. Such a 'blocking high' may be extremely persistent and have a major influence on the weather in the surrounding region. Generally, the Polar Front lies closer to the pole than normal. Any depressions are steered round the high, following tracks that are either further south or north than usual. In winter, a blocking high over Scandinavia, for example, may bring bitter easterly winds over western Europe and, at the same time, force any depressions to take a more southerly track, bringing unusual wet and windy weather to the Iberian Peninsula and the Mediterranean area.

Cross references

Convergence and divergence p.27
Warm and cold highs p.27

BELOW A strong ridge of high pressure, extending across western Europe from north Africa, is acting as a barrier to the normal eastward motion of depressions, and has forced the Polar Front to stall over Britain and the eastern Atlantic. *(Eumetsat)*

PART TWO

Weather processes and precipitation

OPPOSITE Jet-stream clouds illuminated by the setting Sun. The motion of the clouds from the northwest (right to left) indicates that the jet stream lies ahead of the warm front of an approaching depression, still some distance away to the west. *(Author)*

Chapter 5

Weather processes

Water's different phases

Water is an unusual substance in that it may exist in all of its three different forms (phases) at temperatures commonly found on Earth. The transitions between these three states (solid ice, liquid water and gaseous water vapour), and the energy that is exchanged, are involved in many different meteorological processes.

One concept that many people have difficulty in understanding is that of latent heat. It is perhaps best explained by considering how ice turns into liquid water, and liquid water into water vapour. If a piece of ice is raised to melting point (0°C), it remains at that temperature, even with the addition of heat, until all has melted. Similarly, when the temperature of liquid water is raised to boiling point (100°C), it stays at that temperature until all the water has been evaporated into water vapour. In both cases the additional heat does not raise the temperature, but causes the ice or water to change into another phase. Heat is removed from the environment when ice melts into liquid water and when liquid water, in turn, evaporates into water vapour. This 'hidden' or 'invisible' heat is known as latent heat. In the reverse processes, heat must be released into the environment when water condenses into liquid water, and when liquid water freezes into ice.

BELOW The phase diagram of water (H_2O), showing the changes with increasing pressure (in Pascals) and temperature (in Kelvin). S = solid; L = liquid; V = vapour. TP is the triple point at which the three phases are in equilibrium. *(Ian Moores)*

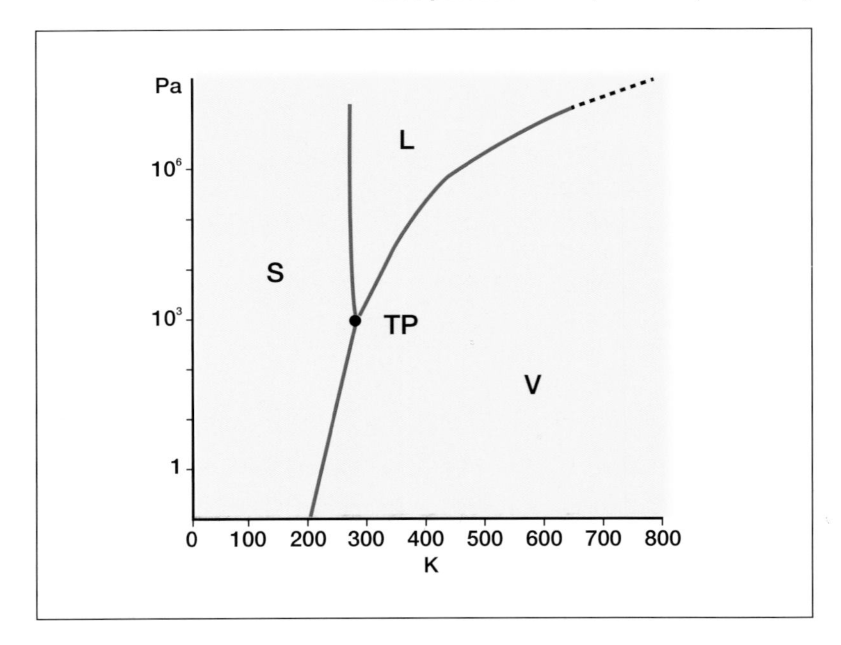

Water is also able to change directly from the solid phase into vapour, and in the reverse process, from vapour into ice. For many years both of these processes have been known as 'sublimation', but in more recent times there has been a tendency, to avoid confusion, to call the change from vapour to solid 'deposition'. Both of these processes occur naturally in the atmosphere. Indeed, the most effective way to dry wet washing is to hang it out in sub-zero temperatures in bright sunlight. The washing initially freezes, and goes as stiff as a board, but then the ice turns directly into water vapour (sublimes), leaving the washing completely dry. Overnight frost crystals may similarly sublime into water vapour when exposed to sunlight, without going through the liquid phase. The opposite process, deposition, occurs both in the tops of clouds and on the ground, causing the growth of ice crystals in both cases. Sublimation removes heat from the environment, and deposition releases latent heat into the surroundings.

The release of latent heat when water freezes into ice is utilised by fruit growers and others to protect their crops from mild frosts. When frost is forecast, trees and bushes are sprayed with water. When the temperature drops overnight, the water droplets freeze, but the latent heat that they release helps to keep the buds or fruit above freezing.

Saturation and humidity

Humidity is a measure of the amount of water vapour in a parcel of air. Consider a closed vessel, partially filled with water and with dry air. All the molecules (of both the water and the gases) are in constant motion, and they show a range of velocities. Some of the molecules in the liquid are moving fast enough to break free into the air: to evaporate and become water vapour. However, as soon as the air contains some water vapour, some of the molecules are moving slowly and enter the water: they condense from water vapour into liquid water. Eventually there will be a balance between the number of molecules leaving the liquid and those entering it. The air has become saturated with water vapour. Its humidity is 100%.

If the container is now heated, the speed of all the molecules increases. More molecules are able to leave the liquid and enter the air, until eventually a new equilibrium is established. The air has become saturated with a greater quantity of water vapour molecules. The reverse will also apply: if the container is cooled the speeds will decrease, and yet another equilibrium will be established with fewer water vapour molecules in the air. (There will always be a few molecules in the air, no matter how low the temperature – unless one were able to take the temperature down to 0K (-273°C), absolute zero, at which all molecular motion ceases.)

The number of water molecules in the air (its humidity) is purely dependent on the temperature, and air is saturated when any additional water vapour molecules would condense immediately if the temperature were to drop. The temperature at which this would happen is known as the dew point. If the container were open to its surroundings and a stream of dry air flowed across it, more water would evaporate, and the additional molecules would be carried away by the stream of air. The fresh air continuously coming into contact with the water would not become saturated. Its humidity would be less than 100%.

In the atmosphere, when saturated air is cooled cloud droplets are created, but these actually require solid particles on which to form. If air is completely free from such particles – a condition that does sometimes occur in the actual atmosphere – the air may be cooled to an even lower temperature than the dew point, when it is said to be super saturated.

The amount of water vapour in a parcel of air is normally measured in terms of the ratio between the mass of the water vapour present and the mass of the total parcel, and is known as the specific humidity. It is normally expressed as the number of grams of water vapour per kilogram of air. The specific humidity of a parcel of air thus rises with a rise in temperature as shown in this table for saturated air:

Temperature	***Specific humidity (g/kg)***
0°C (32°F)	3.0
10°C (50°F)	~7.0
20°C (68°F)	~14.0
30°C (86°F)	26.0

In common usage, the term 'humidity' is taken to be the relative humidity, that is the percentage of water vapour within a parcel of air, relative to the maximum amount that it would contain at saturation. For example, a parcel of air at 20°C (where the saturated ratio would be 14.0g/kg) with a specific humidity of 7.0g/kg would have a relative humidity of 50%. However, relative humidity is of little concern to meteorologists, because the same percentage might be obtained from many other parcels of air at different temperatures. A far more useful concept is absolute humidity, which is the actual mass of water vapour in the air per unit mass of air, usually stated in grams per cubic metre of air.

The standard method of obtaining this information is to take the temperatures shown by two thermometers: a dry-bulb thermometer, simply exposed to the air, and a wet-bulb thermometer, where the bulb has a cover kept permanently moist by distilled water supplied by a wick. These will normally show different temperatures, depending on the humidity of the air. From the two temperatures, and using a standard set of tables, it is possible to determine the absolute humidity (and, of course, the relative humidity, if required). Knowledge of the two temperatures also enables one to

Cross references

Lapse rates p.51
Measuring humidity p.153
Measuring temperatures p.152

Cross references

Composition of air p.14
Supercell storms p.116
Thunderstorms p.114
Tornadoes p.121

determine the dew point temperature – the temperature to which the air must cool before it reaches saturation and condensation begins, and dew is deposited on the ground. It will also be the temperature to which the air must cool for cloud particles to condense. Determining where (at what altitude) this will occur requires a knowledge of the lapse rate – the rate at which temperature changes with height – and this will be discussed shortly.

Knowledge of the dew point is extremely useful, because the value of the dew point in the afternoon provides an indication of the minimum temperature that is likely to prevail during the night. With the fall in temperature after sunset down to the dew point, the release of latent heat as the water vapour condenses tends to offset the fall in temperature. An afternoon air temperature of 30°C, accompanied by a dew point temperature of 22°C, for example, suggests that night-time temperatures will be in the low 20s Celsius, but with an extremely uncomfortable humidity of 100%.

The density of humid air

A common misunderstanding is that humid air 'must' be heavier than dry air – 'because it holds more water'. But this is completely wrong for various reasons. Air does not 'hold' water. It is not like a porous brick or a sponge, with a fixed (or even flexible) structure with cavities that take up water. A wet sponge does indeed weigh more than a dry one. Air, however, is a mixture of gases, and there is a fundamental law (known as Avogadro's Law) that governs the behaviour of gases: for a given volume at a constant temperature and pressure, the overall number of molecules remains constant, no matter which gas or gases are present.

Based on this law, if we add a molecule of water vapour – which may be regarded as simply another form of gas – to our given volume, another molecule of gas is lost. If we consider the different atomic weights of the various molecules, we soon see what happens with a change in humidity.

Very approximately, ignoring lesser components, we may say that air consists of 80% nitrogen, and 20% oxygen: a ratio of 4:1. The atomic weights of N_2 and O_2 molecules are 28 and 32, respectively. (Note that we are talking about molecules, each consisting of two atoms of the gas. A nitrogen atom has an atomic weight of 14, and one of oxygen 16.) A molecule of water (H_2O) has an atomic weight of 18. If we add 10 molecules of water (total atomic weight 180), 8 molecules of nitrogen (total atomic weight 224) and 2 molecules of oxygen (total atomic weight 64) must be displaced. So the atomic weight gained is 180, against a loss of 288. The overall change in atomic weight is thus a loss of 108.

The important point about this is that although two parcels of air may be at exactly the same temperature and pressure, the more humid one has a lower density, and thus tends to rise above the drier parcel. Such a situation may occur along what it known as a 'dryline', separating dry from moist air, and which is often the source of energy for major thunderstorms, including supercell storms which may, in turn, give rise to tornadoes.

CHANGES WHEN HUMIDITY INCREASES

Gas	Atomic weight	Molecules added or lost	Atomic weight added or lost
Water (H2O) introduced	18	+10	180 increase
Nitrogen (N2) displaced	28	-8	224 decrease
Oxygen (O2) displaced	32	-2	64 decrease
Overall atomic weight lost (N2 + O2)	288		
Overall atomic weight gained (H2O)			180
Total change			108 decrease

Stability and instability

As mentioned earlier, pressure decreases with height, and as a consequence any parcel of air that is displaced upwards (by whatever cause) will expand and its temperature will decrease. If, after such an upward displacement, the parcel of air is warmer (and thus less dense) than its surroundings, its buoyancy will cause it to continue to rise, until its temperature eventually becomes the same as its environment. If, after a displacement from its original level, a parcel continues to move in the same direction, the surrounding atmosphere is said to be unstable. Conversely, if after an initial

RIGHT If the wind causes air to rise over high ground, but at its peak the parcel is cooler than its surroundings (the environment) it will return to its original level and temperature on the downwind side. The air is stable. (The levels and temperatures are illustrative, rather than actual measurements.) *(Ian Moores)*

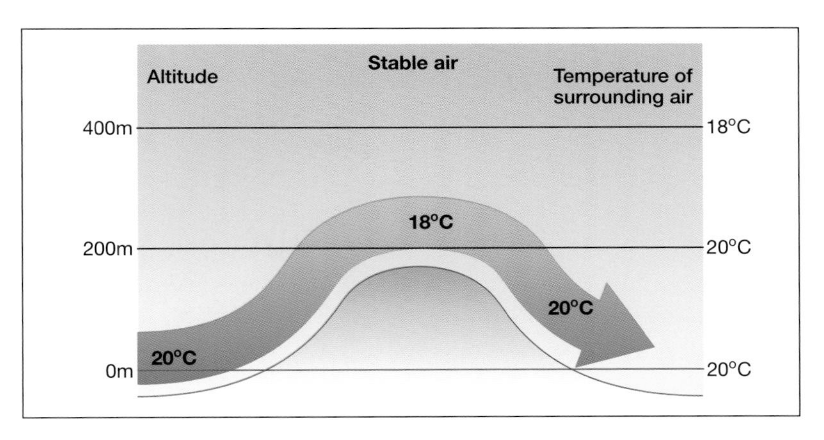

displacement upwards the parcel is cooler (and thus denser) than its surroundings, its buoyancy will tend to return it to its original position. The atmosphere is then said to be stable. It should be noted that the buoyancy of a parcel of air tends to resist any movement (upwards or downwards), and the initial displacement must be forced by some external means.

Lapse rates

As we have seen, the average rate of decline in temperature with height (the lapse rate) is approximately 6.5 deg.C per kilometre up to the tropopause. However, the actual rate may vary considerably. Provided condensation does not take place, a parcel of air that rises in the atmosphere will expand and cool at the rate of approximately 10 deg.C per kilometre (actually 9.767 deg.C km^{-1}). This value is known as the dry adiabatic lapse rate (DALR). ('Adiabatic' means that the parcel of air does not exchange any heat with its surroundings.) The parcel of air will warm at the same rate during descent.

The lapse rate of the surroundings of the parcel of air is all-important. This is the environmental lapse rate (ELR), the actual rate at which temperature changes with height. If the ELR is greater than the DALR, so that the surroundings are always cooler than the rising parcel, even though the latter's temperature is decreasing, the parcel will continue to rise, and the air surrounding the parcel (the environment) is unstable for the ascent or descent of dry air. Conversely, if the ELR is less than the DALR, the surroundings are stable. In practice the ELR is determined by measurements made by a radiosonde, a balloon-borne instrumented package, with ascents made twice a day.

If condensation does occur – the parcel of air having cooled below the dew point – the release of latent heat will reduce the lapse rate to what is known as the saturated adiabatic lapse rate (SALR). Depending on the exact circumstances this will lie between 4 and 7 deg.C per kilometre, and this is the rate that would be measured within a cloud. The SALR has a low value at high temperatures, but increases towards the DALR value at low temperatures. (It becomes approximately equal to the DALR at -40°C.) The reason for the variation in rate is that at high temperatures there is more water vapour able to condense, whereas at low temperatures there are few molecules and thus little latent heat that can be released.

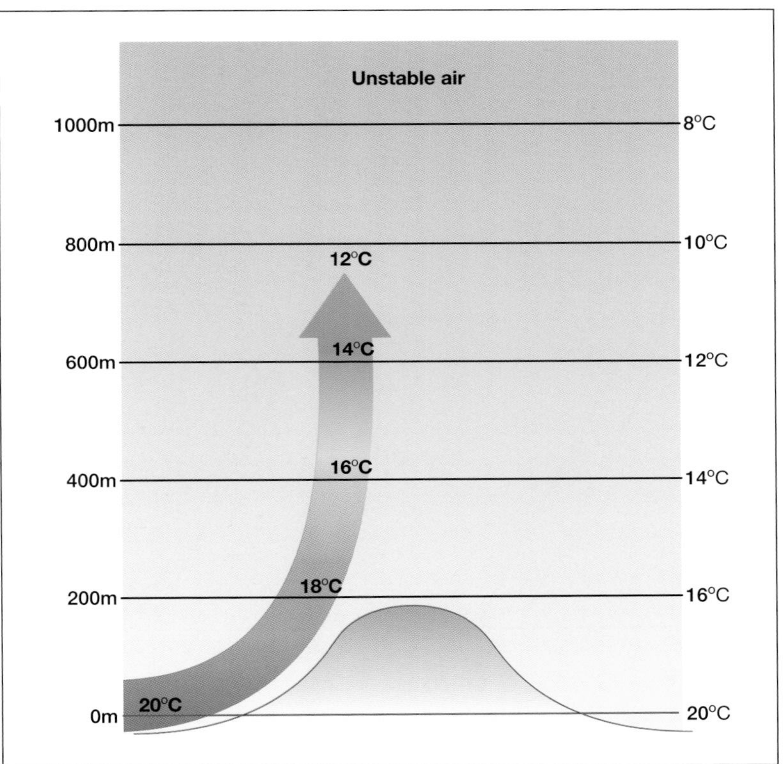

ABOVE When air, despite cooling on rising, still remains cooler than its environment it will continue to rise. (The levels and temperatures are illustrative, rather than actual measurements.) *(Ian Moores)*

Cross references

Dew point p.164
Radiosondes p.145–6

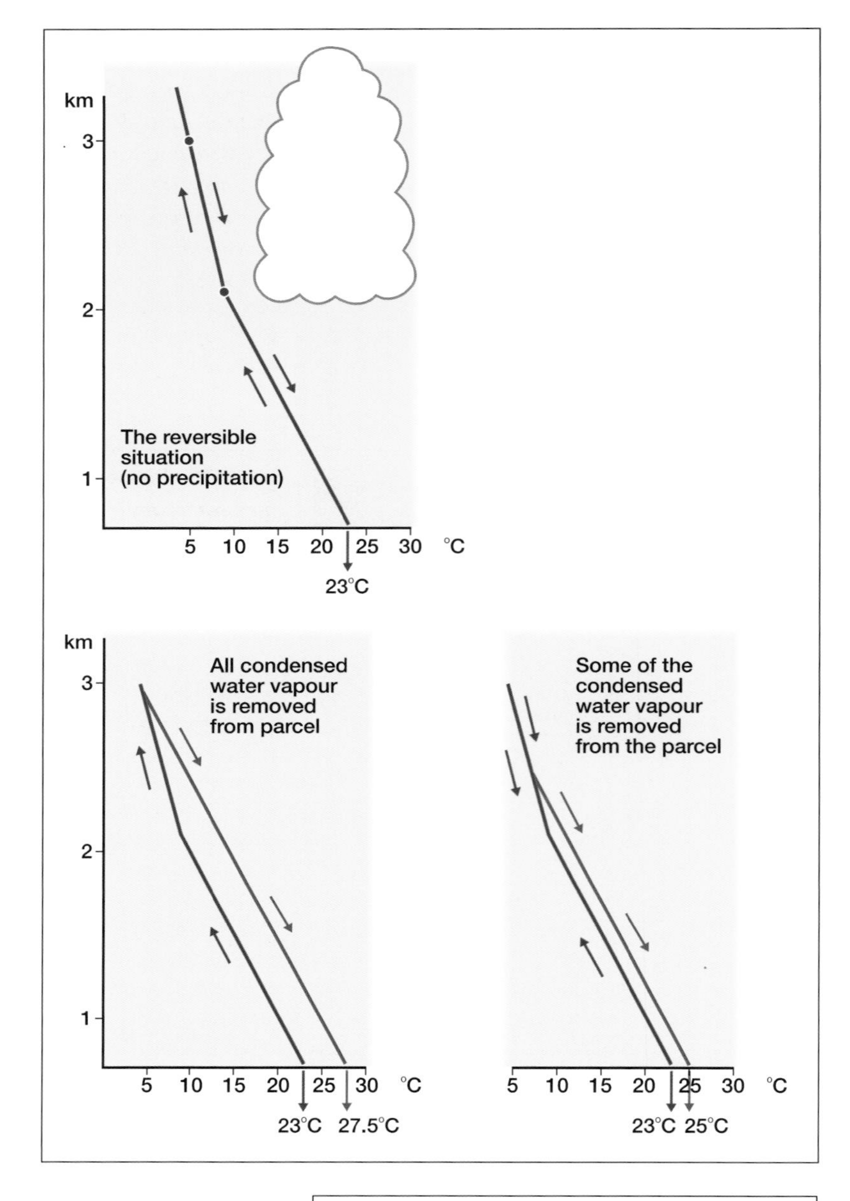

LEFT Idealised lapse rates for humid air: (a) the reversible situation (no precipitation); (b) all the condensed water vapour is removed from the parcel; (c) some of the condensed water vapour is removed from the parcel. Note how the temperatures increase when the parcel returns to its orginal level in (b) and (c). *(Ian Moores)*

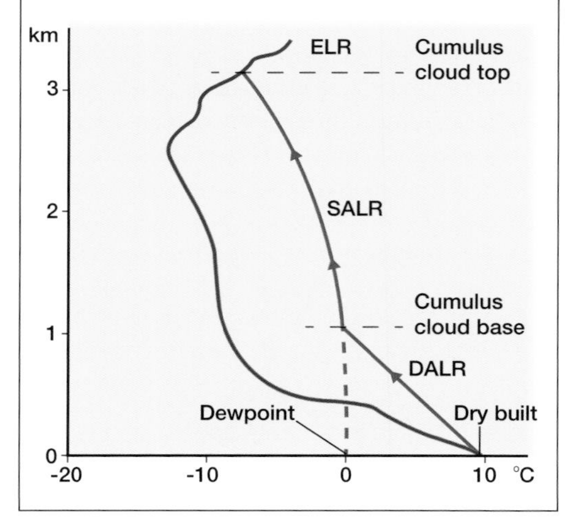

RIGHT An example of lapse rates that might actually occur in the atmosphere and how they affect the growth of Cumulus clouds. *(Ian Moores)*

If all the condensation (or freezing) products remain within the parcel, it will warm at the SALR rate on descent, so that when it reaches its original level it will be at the same temperature. If, however, any products fall out as precipitation (such as rain, snow or hail), the lapse rate will assume a value (known as a pseudo-adiabatic lapse rate) intermediate between the DALR and SALR. Now, when it reaches its initial level, it will be warmer than when it started to ascend. The exact temperature will depend upon whether all, or only a portion, of the condensed water vapour has fallen out of the parcel.

To summarise: all the time a parcel of air is warmer than its surroundings, it will continue to rise at either the DALR or, if condensation has occurred, at the SALR. Similarly, a parcel that is cooler than its environment will sink and warm at the appropriate rate.

Condensation and freezing

Under normal circumstances (unless it is extremely supersaturated) water cannot condense on its own. It requires the presence of suitable condensation nuclei, or for it to come into contact with a free water or ice surface. Condensation nuclei are usually present in large numbers. They include particles produced by combustion such as ammonium sulphate or sulphuric acid droplets; salt crystals (produced in abundance by ocean spray); various soil particles; and even certain bacteria.

In a similar fashion, water will not freeze in the absence of suitable freezing nuclei, unless it is highly supercooled (to about -40°C), or unless it comes into contact with an existing ice surface. In the presence of suitable nuclei water will freeze, but generally this is at a temperature that is below the nominal freezing point of 0°C. It has been found that

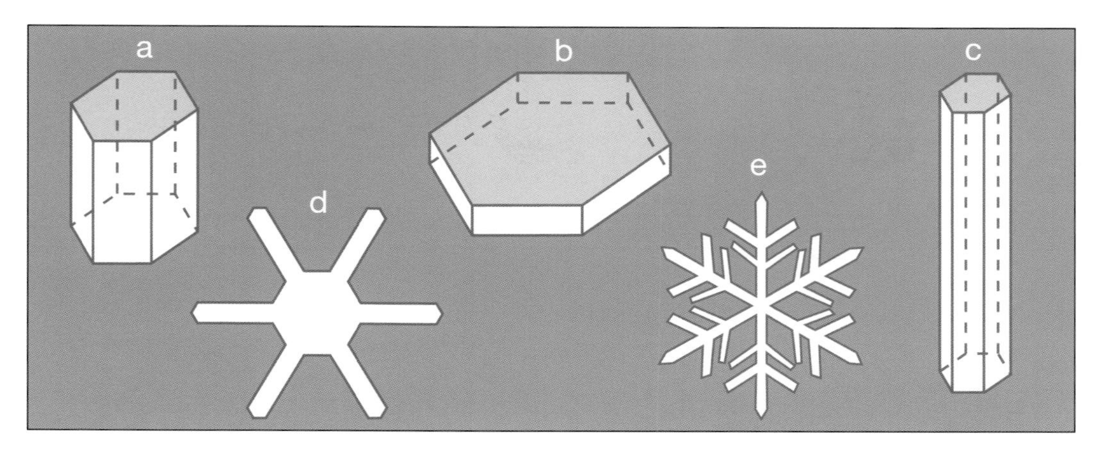

LEFT The various types of ice crystal that form at different temperatures and saturations: (a) hexagonal prism; (b) hexagonal plate; (c) hexagonal needle; (d) stellar crystal; (e) dendritic crystal. *(Ian Moores)*

the most efficient nuclei promote freezing at temperatures between -10°C and -15°C. In general, the nuclei have been found to be certain clay minerals and other insoluble particles. When the freezing process takes place in clouds it is known as glaciation. It is particularly associated with Cumulonimbus.

In the case of noctilucent clouds, which are known to consist of ice particles, but which are so high (~85km) that it is difficult to envisage any mechanisms by which either water vapour or nuclei could be injected to such an altitude, it is considered most likely that the water has been derived from interplanetary space (*ie* ultimately from comets), and that the nuclei are either micrometeorites (also from cometary material), or clusters of ions that have been created by cosmic rays from interstellar space.

The exact shape of ice crystals that are formed in the atmosphere has been found to be closely dependent on the exact temperature and saturation regime. The most common forms are in the shape of prisms: flat, hexagonal plates; hexagonal needles (somewhat resembling pencils in shape); and shorter, wider, hexagonal columns, including hollow hexagonal columns. The multi-branched crystals popularly called 'snowflakes' are known technically as dendritic crystals, and those that have six simple, unbranched arms are stellar crystals. The prismatic forms may have hexagonal, pyramidal (*ie* pointed) ends.

The exact shape of the ice crystals precisely determines the location, colour, brightness and shape of the many different halo phenomena. Dozens of different arcs and points of light are known, created by the refraction of light through the crystals, or by reflection from their flat faces.

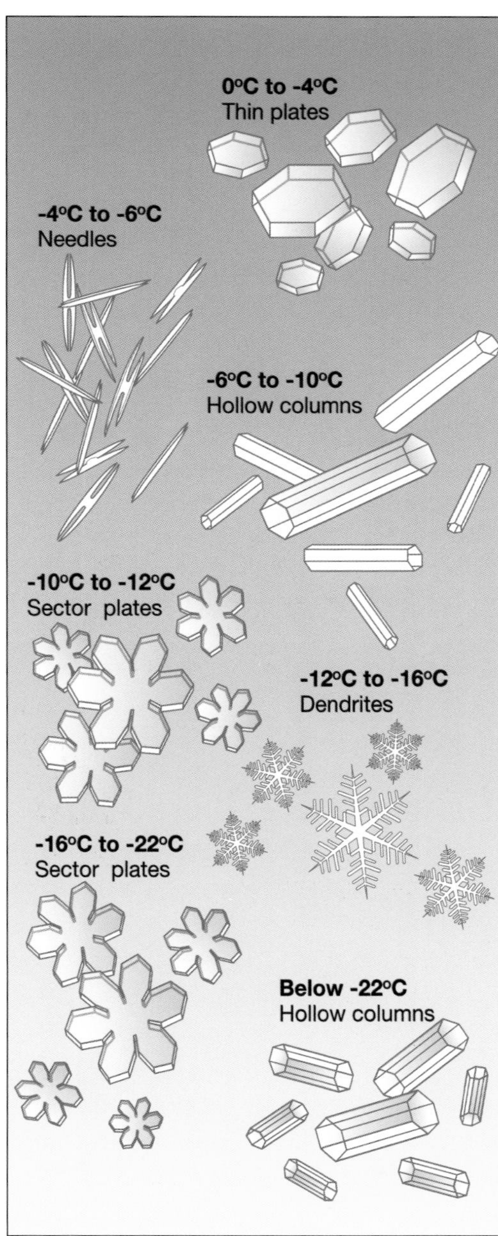

LEFT The temperature and availability of water vapour determine the shape of the ice crystals that form, shown here for various temperature ranges. *(Ian Moores)*

Cross references

Haloes p.135
Noctilucent clouds p.82

Chapter 6

Clouds

Cross references

Accessory clouds p.57
Cloud genera p.55
Cloud species p.56
Cloud varieties p.56
Latin terms p.55
Supplementary cloud features p.57

Cloud classification

Being able to recognise the different types of cloud and other phenomena in the atmosphere is extremely helpful in understanding the weather, both the weather that is occurring at the time and what is likely to happen in the future.

For some strange reason, many people find it hard to recognise clouds, but it is not really difficult. There are ten basic types, and they are quite easy to tell apart. Once you have some idea of these you will start to see differences in their individual shapes or nature, until eventually naming the different forms of cloud will become second nature, and a quick glance at the sky will tell you what is happening and give an idea of forthcoming weather.

Like plants and animals, the basic cloud forms and variants are known by Latin names. Cloud classification was introduced by Luke Howard in 1802, in a talk entitled 'On the Modifications of Clouds', and in a book that was published the following year. The classification was so successful that some of Howard's terms remain in use to this day.

Also as with plants and animals, clouds were originally identified by two Latin terms: the type, known as the genus (pl. genera), and the species. Later a third term, 'variety', was introduced to describe different aspects of the structure and transparency of the particular forms. All these terms correspond to the visual appearance of clouds. In addition, there are accessory clouds that occur in association with one of the main cloud types, and supplementary features that describe specific structures.

This may sound a complicated system, and indeed the overall scheme does allow for the detailed description of a whole range of cloud forms; however, the tables given should enable you to quickly find out the meaning of particular terms or help you decide what a particular cloud or feature should be called. The individual cloud forms and features, and the distinctive characteristics that enable you to recognise them, are described in detail on later pages. With a growth in knowledge of how clouds form and evolve, the World Meteorological Organization introduced a more complex scheme and special symbols for use by official observers, based on cloud altitude and the specific way in which the clouds are developing (or have developed). Details are given later, but for recognition purposes it is best to use the simpler classification.

To summarise, we have:

■ Genus	Overall form
■ Species	Shape and structure
■ Variety	Transparency and arrangement of elements
■ Accessory clouds	Forms that occur only with certain types
■ Supplementary features	Clouds with a distinctive appearance

BELOW **A sky of mixed clouds: high Cirrus with low-level Cumulus and (in the distance) Cumulonimbus clouds.** *(Author)*

Latin and clouds

The names of the different cloud genera, species and varieties (and the accessory clouds and supplementary terms) are derived from Latin. Although the exact meanings and spellings have changed with time, most are related to classical Latin words. One term, 'castellanus', which has become the standard term, really means anything related to a castle. It is thought to have been introduced accidentally into an edition of the *International Cloud Atlas* through a dictionary error, instead of 'castellatus', which means 'crenellated' or 'turreted' – which would more accurately describe the structure of the clouds.

Of Luke Howard's original Latin names, three remain in use and are particularly helpful in subdividing the main types of cloud into groups. These are:

cumulus ('mass' or 'heap')
cumuliform or heaped clouds
stratus ('layer')
stratiform or layer clouds
cirrus ('curl', 'wisp', 'tuft')
cirriform or hair-like clouds

The *International Cloud Atlas*, which is the recognised reference work, published by the World Meteorological Organization, actually divides clouds into just cumuliform and stratiform classes, based on overall appearance. On the basis of the way in which clouds are formed, however – which we shall discuss later – an additional group, 'cirriform' clouds, is occasionally a useful classification, consisting of ice-crystal, rather than water-droplet, clouds.

Cumuliform clouds
Cumulus
Stratocumulus
Altocumulus
Cirrocumulus

The last three of these – Stratocumulus, Altocumulus and Cirrocumulus – usually occur in layers, which are often extremely extensive, so they may be considered to fall into both the cumuliform and stratiform groups.

Stratiform clouds
Stratus
Nimbostratus
Altostratus
Cirrostratus

Cirriform clouds
Cirrus
Cirrostratus
Cirrocumulus

ABOVE Typical Cumulus clouds, photographed fairly early in the day as the land was just beginning to heat up. *(Author)*

Genera

The ten main genera are groups of cloud forms. These forms are specific to a single type: in other words, clouds cannot belong to more than one genus at any one time. On occasion, however, one genus may mutate into another (Altostratus may turn into Nimbostratus, for example). It is usual to write the names of the various genera (and their two-letter abbreviations) with an initial capital letter, as for example, Cumulus (Cu). Specific symbols are used to designate each genus on synoptic charts.

CLOUD GENERA

The ten basic cloud types

Genus	Abbr.	Symbol	Description	Page
Altocumulus	Ac		Heaps or rolls of cloud, with distinct darker shading, and clear gaps between them, in a layer at middle levels.	64
Altostratus	As		Sheet of featureless, white or grey cloud at middle levels.	66
Cirrocumulus	Cc		Tiny heaps of cloud without shading, with clear gaps, in a layer at high levels.	62
Cirrostratus	Cs		Essentially featureless sheet of thin cloud at high levels.	63
Cirrus	Ci		Fibrous wisps of cloud at high levels.	60
Cumulonimbus	Cb		Large, towering cloud that extends to great heights, with ragged base and heavy precipitation.	73
Cumulus	Cu		Rounded heaps of cloud at low levels.	70
Nimbostratus	Ns		Dark grey cloud at middle levels, frequently extending down towards the surface, and giving prolonged precipitation.	67
Stratocumulus	Sc		Heaps or rolls of cloud at low levels, with distinct gaps and heavy shading.	68
Stratus	St		Essentially featureless, grey layer cloud at low level.	69

CLOUD SPECIES

Fourteen terms describing cloud shape and structure

Species	Abbr.	Description	Genera	Page
calvus	cal	The tops of rising cloud cells lose their hard appearance and become smooth.	Cb	74
capillatus	cap	The tops of rising cloud cells become distinctly fibrous or striated; some obvious Cirrus may appear.	Cb	74
castellanus	cas	Distinct turrets rise from an extended base or line of cloud.	Sc, Ac, Cc, Ci	65
congestus	con	Great vertical extent; the clouds are obviously growing vigorously, with hard, 'cauliflower-like' tops.	Cu	72
fibratus	fib	Fibrous appearance, normally fairly straight or uniformly curved; no distinct hooks.	Ci, Cs	63
floccus	flo	Individual tufts of cloud, often with ragged bases, sometimes with distinct virga.	Ac, Cc, Ci	65
fractus	fra	Broken cloud with ragged edges and base.	Cu, St	70,71
humilis	hum	Cloud with a restricted vertical extent; length much greater than height.	Cu	71
lenticularis	len	Lens- or almond-shaped clouds, stationary in the sky.	Sc, Ac, Cc	64
mediocris	med	Cloud of moderate vertical extent, growing upwards.	Cu	71
nebulosus	neb	Featureless sheet of cloud, with no apparent structure.	St, Cs	69
spissatus	spi	Dense cloud, appearing grey when viewed towards the Sun.	Ci	61
stratiformis	str	Extensive sheet or layer.	Sc, Ac, Cc	64
uncinus	unc	Distinctly hooked, often without a visible generating head.	Ci	60

Species

The various terms for species describe observed differences and specific peculiarities of clouds. Such differences, mainly in their internal structure, have meant that nearly all of the ten genera (except Altostratus and Nimbostratus) have been subdivided into particular species. As with the names of genera, the species descriptions are mutually exclusive. A cloud that belongs to a specific genus may be described as being of just one particular species. However, several species are commonly found in different genera. An example is the species designation 'lenticularis', meaning lens-, almond- or lentil-shaped. This species may be applied to Cirrocumulus, Altocumulus, and Stratocumulus. Species terms have three-letter abbreviations.

Varieties

The various varieties describe very specific, observable characteristics that clouds may exhibit. These characteristics are generally related to the clouds' greater or lesser transparency, or to the way in which the cloud elements are arranged. Prominent wave structure in clouds, for example, would be classified with the variety 'undulatus' (*ie* undulating), and a dense layer of cloud that completely blocks the light of the Sun as 'opacus' (opaque).

As with species, variety descriptions may be applied to more than one genera. Unlike species, however, any one cloud may exhibit features that are characteristic of more than one variety. A cloud could be both undulating (undulatus) and opaque (opacus), for example. The terms for varieties have two-letter abbreviations.

RIGHT Very high Altocumulus stratiformis, which could almost be classified as Cirrocumulus. *(Author)*

FAR RIGHT Stratocumulus perlucidus duplicatus with wide gaps between the individual cloud elements. *(Author)*

ABOVE The accessory cloud, pannus, seen beneath Nimbostratus, photographed during a break in prolonged rainfall. *(Author)*

Accessory clouds

There are three specific forms of cloud (known as accessory clouds) that occur solely in conjunction with one of the ten main genera, and under no other circumstances. As with varieties, more than one of these accessory clouds may be present at any given time. They have three-letter abbreviations.

Supplementary features

There are six particular and distinctive features that may be exhibited by individual genera and species. Some of these features are common, others are quite rare. They are given three-letter abbreviations.

BELOW A large, clearly defined anvil cloud – Cumulonimbus incus – photographed at sunset. A second anvil is also visible in the background. *(Claudia Hinz)*

CLOUD VARIETIES

Nine terms describing cloud transparency or the arrangement of cloud elements

Variety	Abbr.	Description	Genera	Page
duplicatus	du	Two or more layers.	Sc, Ac, Ac, Cc, Cs	56
intortus	in	Tangled or irregularly curved.	Ci	61
lacunosus	la	Thin cloud with regularly spaced holes, appearing like a net.	Ac, Cc, Sc	62
opacus	op	Thick cloud that completely hides the Sun or Moon.	St, Sc, Ac, As	69
perlucidus	pe	Extensive layer with gaps, through which blue sky, the Sun or Moon are visible.	Sc, Ac	64,90
radiatus	ra	Appearing to radiate from one point in the sky.	Cu, Sc, Ac, As, Ci	89
translucidus	tr	Translucent cloud, through which the position of the Sun or Moon is readily visible.	St, Sc, Ac, As	66
undulatus	un	Layer or patch of cloud with distinct undulations.	St, Sc, Ac, As, Cc, Cs	63,68
vertebratus	ve	Lines of cloud looking like ribs, vertebrae or fish bones.	Ci	N/A

ACCESSORY CLOUDS

Three forms occurring only with one of the ten main genera

Name	Abbr.	Description	Genera	Page
pannus	pan	Ragged shreds of cloud beneath the main cloud body.	Cu, Cb, As, Ns	74
pileus	pil	Hood or cap of cloud above a rising cloud cell.	Cu, Cb	75
velum	vel	Thin, extensive sheet of cloud, through which the most vigorous cells may penetrate.	Cu, Cb	75

SUPPLEMENTARY FEATURES

Six particular forms that particular genera or species may exhibit

Feature	Abbr.	Description	Genera	Page
arcus	arc	Arch or roll of cloud.	Cb, Cu	75
incus	inc	Anvil cloud.	Cb	76
mamma	mam	Bulges or pouches beneath higher cloud.	Cb, Ci, Cc, Ac, As, Sc	76
praecipitatio	pra	Precipitation that reaches the surface.	Cb, Cu, Ns	77
tuba	tub	Funnel cloud of any type.	Cb, Cu	77
virga	vir	Fallstreaks: trails of precipitation that do not reach the surface.	Ac, As, Cc, Cb, Cu, Ns, Sc, (Ci)	77

Cross references

Cloud altitudes p.58
Synoptic charts p.143
Nacreous clouds p.78
Noctilucent clouds p.82

Additional terms

Two additional terms are sometimes used. These are the suffixes '-genitus' and '-mutatus' – abbreviated '-gen' and '-mut' respectively. These are added to the name of a particular cloud genus, and indicate the type from which a currently visible cloud has been derived.

A portion of a cloud may extend and develop into a different genus. This new form is then classified according to its new type, followed by the name of the genus from which it arose, with the addition of the suffix 'genitus'. For instance, 'Altocumulus cumulogenitus' (Ac cugen) indicates that the predominant cloud is Altocumulus that has been formed from Cumulus, much of which persists.

If the whole or a large portion of a cloud has undergone a major change, it may become a completely new genus. It is then given the name of the new form, followed by the name of the original genus, with the suffix 'mutatus'. An example might be 'Stratus stratocumulomutatus' (St scmut) indicating the presence of Stratus, which previously formed a layer of Stratocumulus.

The tables on the previous spread give brief descriptions of the various features associated with a particular genus, species, variety etc, together with references to where more extensive descriptions and illustrations are given later. The genera with which particular species and varieties may be associated are also shown. However, some varieties occur only with certain species and not with all the clouds of a particular genus. Details of these cases will be given later.

In the tables, the terms are arranged alphabetically for convenience, but it should be noted that clouds are normally arranged in accordance with the altitude at which specific genera occur. This scheme and cloud altitudes will be discussed later.

BELOW Typical ranges of altitude of certain cloud types. *(Ian Moores)*

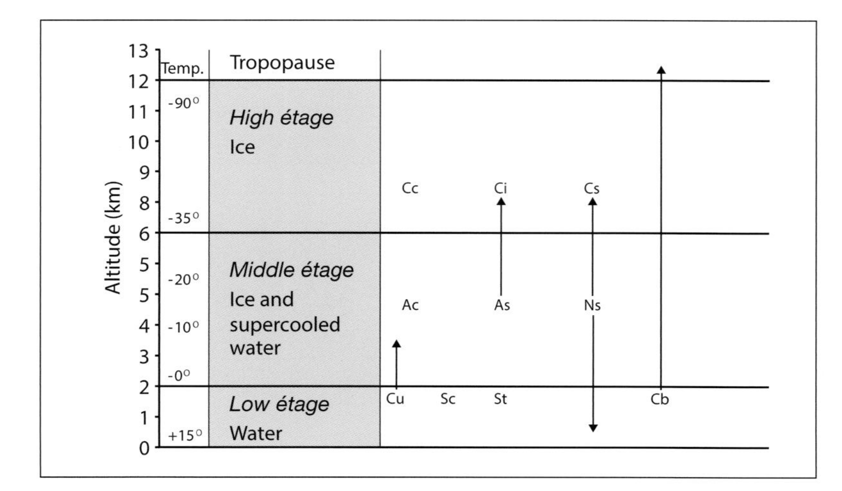

The altitude of clouds

In discussing clouds, a distinction must be made between the terms 'height' and 'altitude'. The height of a cloud is taken to be the vertical distance between the observer (who may be on a hill or mountain) and the cloud concerned. The altitude of a cloud (usually taken to be a point on the base or top) is the vertical distance above mean sea level.

Étages

The altitudes of the main cloud genera have been found by observation to lie between sea level and 18km (60,000ft) in the tropics, approximately 13km (45,000ft) at middle latitudes, and up to 8km (25,000ft) in the polar regions. (This does not apply to the two, relatively rare, high cloud forms: nacreous and noctilucent clouds.)

It has become the accepted convention when discussing clouds for the atmosphere to be divided into three height ranges, known as 'étages' (French: 'level'): high, medium and low. (The designations C_H, C_M, and C_L are used for these levels in observational reports.) The étages are actually defined by the range of levels at which certain specific cloud genera generally occur. The genera used in the definition are:

High
- Cirrus
- Cirrocumulus
- Cirrostratus

Medium
- Altocumulus

Low
- Stratus
- Stratocumulus

The various étages actually tend to overlap and their altitudes vary with latitude. The approximate altitudes are shown in the table.

Altitude ranges of high, medium and low clouds

Étage	*Tropical regions*	*Temperate regions*	*Polar regions*
High	6–18km (20,000–60,000ft)	5–13km (16,000–45,000ft)	3–8km (10,000–25,000ft)
Medium	2–8km (6,500–25,000ft)	2–7km (6,500–23,000ft)	2–4km (6,500–13,000ft)
Low	Surface to 2km (6,500ft)	Surface to 2km (6,500ft)	Surface to 2km (6,500ft)

It will be noted that certain genera are not included among those used to define the three étages. This is because they may occur at a range of altitudes:

- Altostratus is normally regarded as a middle-level cloud, but may extend upwards into the high étage.
- Nimbostratus is likewise usually discussed as a middle-level cloud, but often extends to higher and lower levels.
- Both Cumulus and Cumulonimbus are regarded as low-level clouds because of the altitude of their bases, but they often have great vertical extent, reaching the middle and high étages. Cumulonimbus, in particular, frequently reaches the tropopause.

Without considerable experience or specialised equipment, it is extremely difficult to estimate altitudes accurately, so when learning the different cloud types and features, identification is usually based on the appearance of a particular form of cloud with, perhaps, a knowledge of how it has arisen. When the altitude of a particular cloud is known – perhaps by observing it from an aircraft, cruising at a known altitude – knowledge of the likely étage and the genera found at that level may help to determine the actual genus present.

Another factor that must be borne in mind is that clouds may display one set of features in one part of the sky, and different features elsewhere. A sheet of Altostratus may break up into Altocumulus, for example. Such changes also frequently occur with time. A further complication is that several cloud types may be present at any one time, and that lower clouds may partially hide higher forms, making the latter difficult to identify. Some stratiform clouds, in particular, may exist as distinct layers at more than one level (in the variety known as duplicatus), often making it difficult to decide whether the higher cloud is the same type or whether it belongs to a different class.

Cirrus

Cirrocumulus

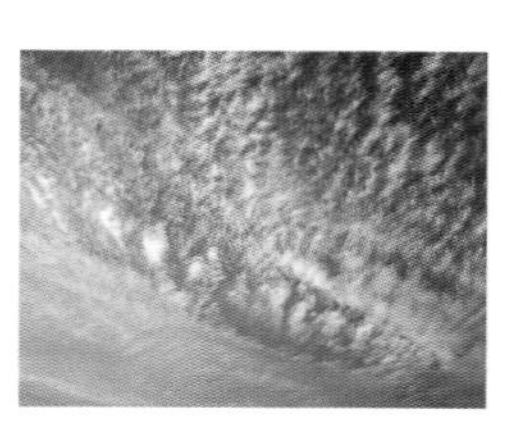

Cirrostratus

Altocumulus

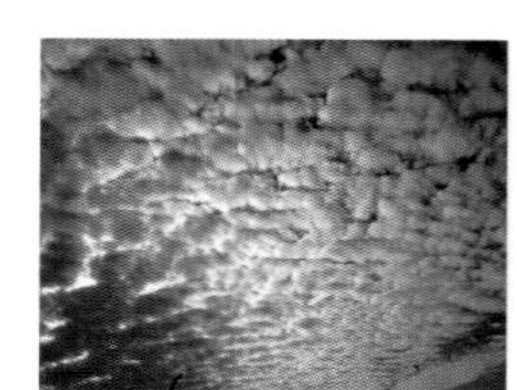

Altostratus

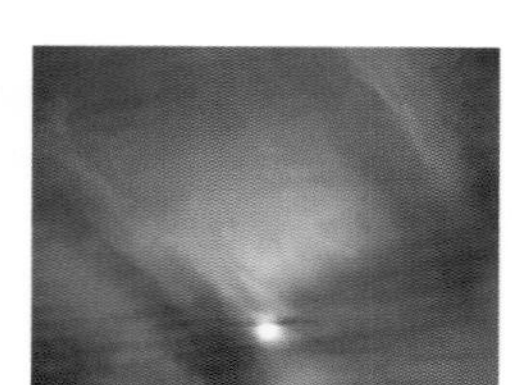

Nimbostratus

Stratus

Stratocumulus

Cumulus

Cumulonimbus

ABOVE AND LEFT
The major cloud types arranged by étage.
(Author)

Cloud forms

In the following descriptions of the various cloud forms, species and varieties are generally described under the main heading for particular genera. Within each genus, the lists of species and varieties are given in a very approximate order of frequency of occurrence. Varieties would normally be given (as abbreviations) following a specific species abbreviation, but in the lists the species are omitted. Not all species or varieties of a genus are described in detail, but some important types are given individual treatment, especially where a particular genus is noted for displaying specific features that may serve as examples for the species or variety in general.

Accessory clouds and supplementary features are listed individually after the main cloud genera, but may also be mentioned in the discussion of individual genera.

High clouds

Cirrus (Ci)

All three genera in the high étage are ice-crystal clouds and, of these, Cirrus is the most familiar, popularly known as 'mare's tails'.

Cirrus always consists of very fine trails or filaments of cloud, which sometimes give it a 'silken' appearance, and the popular name of 'mares' tails'. These filaments may be reasonably straight, curved, with hooks at one end, or seemingly randomly entangled. A frequent form is the hooked shape (in the species Cirrus uncinus), somewhat like a very elongated comma, with a small tuft at the tip. These small tufts of cloud are the generating heads, where the ice crystals are forming, and the streamers following them are actually a form of virga where the slowly descending crystals trail behind the heads that are being carried along by the wind. Occasionally there is very little wind shear and the crystals fall into a layer that is moving uniformly throughout its depth, giving rise to long, almost vertical trails.

Infrequently, when the trails of ice crystals are dense and lie almost horizontally across

RIGHT A fine display of Cirrus, illustrating the change in wind direction, from north-westerly upper winds to south-westerly lower winds, ahead of an advancing depression. *(Author)*

RIGHT Moderately dense Cirrus fibratus is often the precursor of a complete sheet of Cirrostratus. *(Author)*

FAR RIGHT An extensive display of Cirrus uncinus, or hooked cirrus, that covered the whole sky. *(Author)*

RIGHT A dense plume of Cirrus that formed the anvil of a distant Cumulonimbus cloud. *(Author)*

FAR RIGHT A bright parhelion (mock Sun) in a patch of Cirrus. *(Author)*

FAR LEFT Tangled wisps of Cirrus intortus. *(Author)*

LEFT Although appearing white here, this Cirrus spissatus is so dense that it would appear grey when seen against the light. *(Author)*

the sky, they may exhibit the downward bulges known as mamma.

The generating heads are often little more than slightly denser patches of cloud, but occasionally, in the species Cirrus floccus, they may show distinct, small, isolated rounded heads. These heads often form in what is otherwise clear air. Very occasionally they appear as turrets that arise from a lower base (Cirrus castellanus).

Parallel bands of Cirrus (Cirrus radiatus) are reasonably common, and these are particularly associated with jet streams, which are often visible as streaks of Cirrus that are carried along by the high-speed, high-altitude winds. Transverse billows are also frequently seen in jet-stream Cirrus.

When high in the sky, Cirrus clouds appear white, and are generally brighter than other clouds in the sky. When dense patches occur, however, they may be thick enough to seem grey when seen against the light, and one species, Cirrus spissatus, may even be thick enough to block the light of the Sun. Because of their height, Cirrus often remain white at sunset, because they are still fully illuminated by sunlight, when lower clouds have taken on yellow, orange or red tints. Somewhat later, Cirrus clouds themselves become coloured, when lower clouds stand out against them as dark silhouettes.

Cirrus clouds often exhibit optical phenomena – strongly coloured parhelia in particular may be very striking in moderately dense patches, but because the clouds occur as patches, rather than as an extended sheet like Cirrostratus, the larger haloes are rarely seen, although short arcs do sometimes appear.

Cirrus may form from Cirrocumulus, Altocumulus, and occasionally through the break-up of a sheet of Cirrostratus. Very dense plumes of Cirrus are commonly produced by Cumulonimbus (at the Cumulonimbus incus stage), and these plumes often persist long after the parent cloud has dissipated, when they may be known as 'detached anvils'. A similar form of Cirrus often trails behind the cold front of a depression.

Cirrus also commonly arises from aircraft condensation trails (contrails), when freezing conditions occur at flight altitude. Depending on the exact conditions, such trails may be very persistent, spreading to give broad bands of Cirrus across the sky.

Cross references

Contrails p.93
Mamma p.76
Virga p.77

Cirrus species		
Cirrus fibratus	(Ci fib)	Fibrous Cirrus
Cirrus uncinus	(Ci unc)	Hooked Cirrus
Cirrus spissatus	(Ci spi)	Dense Cirrus
Cirrus castellanus	(Ci cas)	Turreted Cirrus
Cirrus floccus	(Ci flo)	Tufted Cirrus
Cirrus varieties		
Cirrus intortus	(Ci in)	Twisted Cirrus
Cirrus radiatus	(Ci ra)	Cirrus in rows
Cirrus vertebratus	(Ci ve)	Cirrus like fish bones
Cirrus duplicatus	(Ci du)	Multiple levels of Cirrus
Cirrus supplementary features		
mamma	(mam)	

RIGHT A sheet of Cirrocumulus stratiformis, grading into a thicker sheet of Cirrostratus in the distance. *(Author)*

FAR RIGHT TOP Part of a very extensive sheet of Cirrocumulus lacunosus. *(Author)*

FAR RIGHT The generating heads in Cirrocumulus producing virga that are forming into a sheet of Cirrostratus. *(Author)*

Cross references

Altocumulus p.64
Coronae p.134
Iridescence p.134
Jet-stream clouds p.94
Measuring angles on the sky p.158
Wave clouds p.84–87

Cirrocumulus (Cc)

Cirrocumulus is often inconspicuous and may be easily overlooked, both because the cloud is thin, and does not greatly affect sunlight, and also because the individual cloud elements are relatively tiny. One criterion that may be used to distinguish Cirrocumulus from the lower Altocumulus is that the cloud elements are less than 1° across, measured at an elevation of 30° from the horizon. The lack of any shading also distinguishes them from Altocumulus elements.

A sheet of Cirrocumulus (Cirrocumulus stratiformis) is far less distinctive than similar layers of Altocumulus and Stratocumulus. This is primarily because – apart from the elements appearing smaller because of their greater height – it is usually very much thinner. (There is, in fact, an empirical relationship between these three cloud genera: the lower the cloud, the larger the individual elements, and the thicker the layer.) Sometimes Cirrocumulus appears as extremely delicate ripples of cloud (in the variety Cirrocumulus undulatus), and these may be so low in contrast that they are difficult to distinguish against the background sky.

In general Cirrocumulus clouds are very thin, and not only allow the position of the Sun or Moon to be seen through them but also do not prevent the light from either of these bodies from casting shadows. The clouds usually consist predominantly of ice particles, although they may also contain supercooled water droplets that normally rapidly freeze into ice particles. The cloud particles are frequently even in size and may give rise (in particular) to coronae and iridescence.

Although Cirrocumulus normally occurs as the species Cirrocumulus stratiformis in an extensive sheet or smaller patch, it may appear as very high, smooth wave clouds (Cirrocumulus lenticularis). Less often, distinct

Cirrocumulus species		
Cirrocumulus stratiformis	(Ci str)	Wide layer of Cirrocumulus
Cirrocumulus lenticularis	(Ci len)	Lenticular Cirrocumulus
Cirrocumulus castellanus	(Ci cas)	Turreted Cirrocumulus
Cirrocumulus floccus	(Ci flo)	Tufted Cirrocumulus
Cirrocumulus varieties		
Cirrocumulus undulatus	(Cc un)	Undulating Cirrocumulus
Cirrocumulus lacunosus	(Cc la)	Cirrocumulus with holes
Cirrocumulus radiatus	(Cc ra)	Cirrocumulus in long lines
Cirrocumulus supplementary features		
virga	(vir)	
mamma	(mam)	

turrets may be seen rising from a level base in the species Cirrocumulus castellanus, or it may occur as separate tufts of cloud (Cirrocumulus floccus). The most uncommon form is possibly the variety Cirrocumulus lacunosus, where the thin layer is punctured by a more or less regular pattern of holes.

Cirrocumulus frequently arises from a layer of Cirrostratus, when shallow convection breaks up the layer into tiny cloudlets, and it may also remain following the decay of Altocumulus. Less frequently it may arise from the thickening and clumping together of Cirrus elements.

Regularly patterned Cirrocumulus in the varieties undulatus and radiatus (together with similar high Altocumulus) are commonly known as 'mackerel skies' after a fancied resemblance to the striped pattern on the fish. These two varieties are often seen in jet-stream Cirrus, although generally not in the form of very regular patterns.

Cirrostratus (Cs)

Cirrostratus is the high-cloud genus that is often overlooked by the general public, even if they have some idea of the different cloud genera. It usually spreads slowly across the sky as a very thin veil. Initially it hardly affects the sunlight, but gradually the sunshine seems to lose some of its heat. It is then that people become aware that the thin cloud has covered the sky. Sometimes Cirrostratus is perfectly featureless – when it is known as Cirrostratus nebulosus – and the only sign that it is present is a milky veil across the sky. Normally, however, it shows distinct signs of a fibrous structure, even if only in isolated patches. It is then described as being the species Cirrostratus fibratus. Cirrostratus is always thin enough for the Sun (or Moon) to be seen through it and to cast shadows.

At times the edge of a sheet of Cirrostratus may be quite sharp, but generally individual wisps of Cirrus gradually become more numerous, until they produce a sheet of Cirrostratus completely or partially covering the sky. This generally occurs with the approach of the warm front of a depression, and the Cirrostratus itself usually gradually thickens and descends towards the surface, eventually becoming a sheet of Altostratus.

Cirrostratus itself may be considered a relatively uninteresting form of cloud, but it is notable for the vast array of halo phenomena that it may produce. These include the 22° and 46° haloes, parhelia (mock suns), the circumzenithal arc, and many other arcs and points of light. (These and other optical phenomena are described later.) These effects are so striking and common that they are diagnostic of the presence of Cirrostratus. However, they may be quite fleeting and are most conspicuous when the sheet of cloud is thin. They disappear as the cloud thickens.

Cirrostratus most often occurs (as mentioned)

LEFT An extensive sheet, rather unusually dense, of Cirrostratus fibratus. *(Author)*

LEFT A typical sheet of Cirrostratus, with a 22° solar halo. *(Author)*

Cross references

Altostratus p.66
Cirrocumulus p.62
Cirrus p.60
Depression p.39
Haloes p.135
Warm front p.39

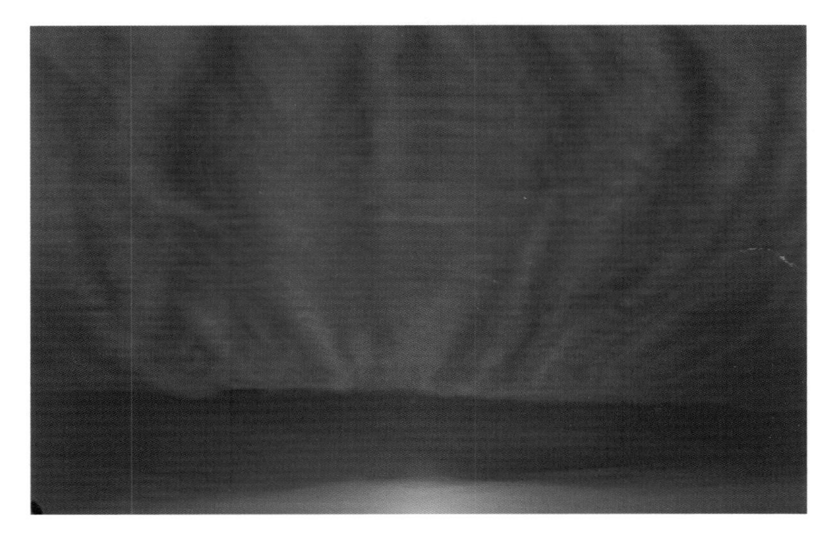

BELOW An extensive sheet of Cirrostratus undulatus, illuminated by the setting Sun. *(Author)*

Cirrostratus species		
Cirrostratus fibratus	(Cs fib)	Fibrous Cirrostratus
Cirrostratus nebulosus	(Cs neb)	Uniform Cirrostratus
Cirrostratus varieties		
Cirrostratus duplicatus	(Cs du)	Multiple layers of Cirrostratus
Cirrostratus undulatus	(Cs un)	Undulating Cirrostratus

Cross references

Cirrocumulus p.62
Coronae p.134
Halo (ice-crystal) phenomena p.135
Iridescence p.134
Measuring angles on the sky p.158
Stability and instability p.50
Stratocumulus p.68
Wave clouds p.84–87

ahead of an approaching warm front, but it may also be produced by the spreading out of a Cirrus plume created by Cumulonimbus. Occasionally it is created by ice crystals that fall from a layer of Cirrocumulus. On rare occasions it may remain when a layer of Altostratus decays.

Medium clouds

Altocumulus (Ac)

Altocumulus has many features in common with the higher Cirrocumulus and the lower Stratocumulus. Like them, it consists of a layer of individual masses of cloud, separated by gaps of clear sky. It may be distinguished from these other genera by the size of the elements, which are less than 5° across, measured at an elevation of 30° from the horizon, and greater than 1° (the size of Cirrocumulus elements). Another distinguishing feature is the amount of shading, which is lighter than the dark shading of Stratocumulus, and greater than Cirrocumulus (which shows no shading at all).

Like these similar genera, Altocumulus normally appears as a sheet of more or less regular individual cloud elements, which may be small rounded masses, flattened 'pancakes' or larger rolls of cloud. All these elements normally show a distinct amount of shading, and blue sky is visible between them. When the elements are close together, the base of the layer often exhibits a lumpy appearance, especially under conditions of low lighting at dawn or dusk.

Sometimes the cloud elements may be in the form of individual tufts (Altocumulus floccus), or there may be turrets of cloud rising up from a more or less continuous base in the species Altocumulus castellanus. Both of these species are an indication of instability at that level. If lower Cumulus clouds reach that level, they often experience rapid growth, swiftly developing into Cumulus congestus or Cumulonimbus, and may lead to heavy showers or thunderstorm activity. Virga are commonly associated with Altocumulus, and when there are rows of cloud the bases sometimes develop distinct mamma, which are usually clearer to see than those that develop beneath Cirrus clouds.

Altocumulus varies greatly in its degree of transparency. It may completely mask the position of the Sun and Moon, or may equally be so thin that their positions may be detected quite easily. It may consist entirely of water droplets (although often in a supercooled state), but is frequently a mixed cloud with both supercooled water droplets and ice crystals. One result of this is that, depending on circumstances, Altocumulus may exhibit optical phenomena produced by water droplets (such as coronae and iridescence) or by ice crystals (such as parhelia and sun pillars), although these halo phenomena are less common.

RIGHT A well-defined sheet of Altocumulus stratiformis perlucidus. *(Author)*

RIGHT Altocumulus lenticularis, generated in the crests of wave motion created by distant hills. *(Author)*

RIGHT A train of Altocumulus lenticularis wave clouds, downwind of low hills off to the right. *(Author)*

Altocumulus, like Cirrocumulus and Stratocumulus, forms either from the uplift of a humid layer of air, or through the break-up of an existing layer (normally Altostratus, but occasionally with the dispersal of deep Nimbostratus, especially in association with the clouds in a depression). Altocumulus frequently appears at the edges of a sheet of Altostratus, and occasionally may form from the thickening of a layer of Cirrocumulus.

Because of their altitude and general structure, and depending on the lighting, Altocumulus (particularly Altocumulus stratiformis covering a large part of the sky) display some of the most dramatic cloud effects. When affected by wind shear, the elements in the variety Altocumulus undulatus may form very striking billows covering the sky.

The species Altocumulus lenticularis may form some of the most dramatic effects ever seen. They are wave clouds that form lens- or almond-shaped clouds that hang motionless in the sky. Depending on the conditions, there may be several humid layers, one above the other, giving rise to a stack of individual clouds, known by the French term 'pile d'assiettes' ('pile of plates'). Close examination of the clouds (perhaps with binoculars) will reveal that they are condensing on the upwind edge, and dissipating downwind as the air passes through them. They remain stationary in the sky all the time that the wind direction and velocity remain constant, but may disappear, or change position, if either factor alters. Again depending on the exact conditions, Altocumulus lenticularis clouds may not appear as individual isolated clouds, but as an elongated trail with wavy upper and lower surfaces that extends downwind from the high ground that is initiating the wave motion.

Although both Cirrocumulus and Stratocumulus occur in the lenticularis variety, Altocumulus lenticularis tend to be more striking, partly because Cirrocumulus tends to be thin (and higher), and thus more difficult to see (unless the observer is high in the mountains), and Stratocumulus lenticularis suffers from the opposite factor: it is often too low for details to be clearly visible, and the dense layer of cloud frequently obscures details of the wave clouds.

LEFT An extensive field of Altocumulus floccus, a sign of instability at height. *(Author)*

LEFT Altocumulus floccus with virga, photographed from the south coast. By the next day, the unstable air had reached Scotland, giving rise to major thunderstorms and even a tornado. *(Author)*

LEFT The turrets of Altocumulus castellanus are a certain indication of instability at the clouds' altitude. *(Author)*

Altocumulus species		
Altocumulus stratiformis	(Ac str)	Wide sheet of Altocumulus
Altocumulus lenticularis	(Ac len)	Lenticular Altocumulus
Altocumulus castellanus	(Ac cas)	Turreted Altocumulus
Altocumulus floccus	(Ac flo)	Tufted Altocumulus
Altocumulus varieties		
Altocumulus translucidus	(Ac tr)	Translucent Altocumulus
Altocumulus perlucidus	(Ac pe)	With gaps Altocumulus
Altocumulus opacus	(Ac op)	Opaque Altocumulus
Altocumulus duplicatus	(Ac du)	Multiple layers of Altocumulus
Altocumulus undulatus	(Ac un)	Undulating Altocumulus
Altocumulus radiatus	(Ac ra)	Altocumulus in rows
Altocumulus lacunosus	(Ac la)	Altocumulus with holes
Altocumulus supplementary features		
virga	(vir)	
mamma	(mam)	

Cross references

Cirrostratus p.63
Coronae p.134
Iridescence p.134
Nimbostratus p.67
Stratocumulus p.68
Stratus p.69

RIGHT Steadily thickening Altostratus ahead of an approaching warm front. *(Author)*

RIGHT Altostratus breaking up following the passage of a frontal system. *(Author)*

RIGHT Altocumulus translucidus clearly showing the position of the Sun, well ahead of an approaching warm front. *(Author)*

RIGHT A sheet of Altostratus, showing some structure, suggesting that it could be classed as Altostratus radiatus. *(Author)*

Altostratus varieties

Altostratus translucidus	(As tr)	Translucent Altostratus
Altostratus opacus	(As op)	Opaque Altostratus
Altostratus duplicatus	(As du)	Multiple layers of Altostratus
Altostratus undulatus	(As un)	Undulating Altostratus
Altostratus radiatus	(As ra)	Altostratus in rows

Altocumulus accessory clouds

pannus	(pan)

Altocumulus supplementary features

virga	(vir)
praecipitatio	(pra)
mamma	(mam)

Altostratus (As)

Altostratus has sometimes been described as an uninteresting and boring cloud. It is a significant feature of an approaching warm front and is generally an extremely extensive and thick layer of cloud, which may be hundreds of kilometres across and thousands of metres thick. At a warm front, it usually develops as Cirrostratus lowers and thickens, and in its initial stages often appears fibrous, but this gradually disappears and the layer becomes essentially featureless. At first, it is the variety Altostratus translucidus and allows the position of the Sun or Moon to be detected through it, although appearing as if seen through ground glass, and with rather diffuse shadows. It then normally becomes opaque (Altostratus opacus) and the Sun no longer casts a shadow.

Altostratus is a mixed cloud, consisting of tiny cloud droplets (usually supercooled), ice crystals and snowflakes. This mixture of different particles means that it rarely shows any optical phenomena, although occasionally the edges of a sheet of Altostratus may consist of just water droplets, and thus display some slight iridescence or partial coronae around the Sun or Moon.

Altostratus usually produces large quantities of precipitation, and this may reach the ground as rain, snow or ice pellets. It also frequently exhibits virga, and these often contribute to the striated appearance when a layer begins to cover the sky. Similarly, ragged pannus accessory clouds often appear just below its base.

Although most commonly observed at temperate latitudes ahead of an approaching warm front, Altostratus also frequently forms from Nimbostratus behind a frontal system. Altostratus may also form, at low latitudes, if vigorous Cumulonimbus clouds or thunderstorm systems spread out at their middle or upper layers. Such Altostratus tends to trail behind the generating system. (Similar Altostratus may sometimes form at middle latitudes, but there Stratocumulus and Stratus are more commonly encountered under similar circumstances.)

As mentioned, there are no species of Altostratus, only five varieties.

Nimbostratus (Ns)

Nimbostratus is the cloud genus that shows the least variation in its appearance. It does not occur as any varied species and, unlike Altostratus, does not even exist in different varieties. It is associated with a single accessory cloud (pannus) and just two supplementary features (praecipitatio and virga).

The most widespread layers of Nimbostratus are produced by the slow uplift of warm, humid air ahead of an advancing warm front, when it usually follows, more or less imperceptibly, from thickening and lowering Altostratus. When it occurs on slow-moving occluded fronts it may produce days of persistent rain or snow, and may thus cause major flooding or record-breaking snowfalls.

Nimbostratus is usually a very deep cloud and, unlike Altostratus, is never thin enough for the position of the Sun or Moon to be seen through it. It may extend upwards, well beyond the nominal middle étage and into the high étage. It also tends to extend downwards, sometimes even reaching ground level. Covering such a wide range of altitudes, it also contains a variety of cloud particles: liquid water droplets; supercooled droplets, snow and ice crystals. It regularly produces heavy, prolonged rain, but when the temperature of the ground is below freezing the rain may freeze on impact to give rise to glaze (black ice). When Nimbostratus advances over a layer of very cold air it may produce exceptionally heavy falls of snow.

Infrequently, Nimbostratus may arise through the thickening of layers of Altocumulus or Stratocumulus. On very rare occasions, Cumulus congestus and Cumulonimbus clouds or thunderstorm systems may sometimes extend and produce a layer of Nimbostratus, but these are always relatively limited in area, unlike the vast sheets of cloud that occur in depressions.

Large Cumulonimbus clouds, multicell storms and supercell storms may cover a large area of the sky and their bases may somewhat resemble Nimbostratus cloud, especially as they may be accompanied (like Ns) by patches of the pannus accessory cloud. Normally, however, it is easy to determine which cloud type is involved from the prevailing conditions, especially if thunder or lightning are observed, or if the precipitation is in the form of hail, when it is safe to conclude that major convective cloud is present. At night-time, if precipitation (rain or snow) is reaching the ground it is conventional to assume that the cloud is Nimbostratus, rather than Altostratus.

Thick Stratus may be confused with Nimbostratus, but the former never gives rise to heavy precipitation, producing only drizzle, tiny snow grains or minute ice crystals.

Nimbostratus accessory clouds	
pannus	(pan)

Nimbostratus supplementary features	
praecipitatio	(pra)
virga	(vir)

Cross references

Altostratus p.66
Frontal systems p.30
Precipitation p.95–99
Stratus p.69
Supercooling p.166
Thunderstorms p114

BELOW A sheet of Nimbostratus beginning to break up following the passage of a frontal system. *(Author)*

LEFT Pannus wisps beneath Nimbostratus, photographed during a break in the otherwise nearly continuous rain. *(Author)*

Cross references

Altocumulus p.64
Cumulonimbus p.73
Cumulus p.70
Measuring angles on the sky p.158
Stability and instability p.50

Low clouds

Stratocumulus (Sc)

Although Stratocumulus is a very varied cloud, which often exhibits a wide range of density and colour, it usually occurs as an extensive layer and consists of individual heaps, masses or 'pancakes' of cloud, with very pronounced shading. It resembles a lower, darker form of Altocumulus, but the cloud elements are much larger, always being greater than 5° across, measured at an elevation of 30° from the horizon.

RIGHT Heavy Stratocumulus clouds over Paris. *(Author)*

RIGHT Stratocumulus with small gaps between the elements, through some of which the Sun is producing crepuscular rays. *(Author)*

RIGHT Distinct billows in a layer of Stratocumulus undulatus. *(Author)*

Stratocumulus species		
Stratocumulus stratiformis	(Sc str)	Wide sheet of Stratocumulus
Stratocumulus lenticularis	(Sc len)	Lenticular Stratocumulus
Stratocumulus castellanus	(Sc cas)	Turreted Stratocumulus
Stratocumulus varieties		
Stratocumulus translucidus	(Sc tr)	Translucent Stratocumulus
Stratocumulus perlucidus	(Sc pe)	Stratocumulus with gaps
Stratocumulus opacus	(Sc op)	Opaque Stratocumulus
Stratocumulus duplicatus	(Sc du)	Multiple layers of Stratocumulus
Stratocumulus undulatus	(Sc un)	Undulating Stratocumulus
Stratocumulus radiatus	(Sc ra)	Stratocumulus in rows
Stratocumulus lacunosus	(Sc la)	Stratocumulus with holes
Stratocumulus supplementary features		
mamma	(mam)	
virga	(vir)	
praecipitatio	(pra)	

Although Stratocumulus always consists of individual cloudlets, which are generally regularly arranged, some of these may be so large that they could be considered patches of Stratus cloud.

Normally the base of a Stratocumulus layer is fairly well defined, lying at a uniform altitude, but if the upper surface is visible (perhaps from an aircraft or mountain above the layer), it may be very uneven. Although Stratocumulus usually occurs in a moderately thin layer, it is often extremely extensive, covering the whole of the sky. (Stratocumulus is the most common cloud over the oceans.) It also often occurs (as Stratocumulus stratiformis duplicatus) at more than one level, although the higher layer (or layers) may be difficult to detect through gaps in the lower cloud.

Stratocumulus primarily forms in two different ways. One applies when a humid layer of stable air rises, being lifted in a depression system, when it is forced to ascend over high ground, or when daytime heating or an increase in wind speed causes a layer of fog or low Stratus to rise. Shallow convection begins in the layer of Stratus that has been produced, often when the top of the cloud radiates heat away to space (even during the daytime), the convective overturning, with cool air subsiding and producing the gaps between the cloud elements.

The second way in which Stratocumulus is often formed is when there is relatively weak convection during the day, and weak thermals produce shallow Cumulus clouds that rise until they encounter an inversion. Here, the clouds spread out sideways, giving rise to the layer of Stratocumulus. Some stronger thermals may break through the inversion and continue rising above the inversion. The strongest may even produce Cumulus congestus or Cumulonimbus clouds.

Restricted areas of Stratocumulus are often found ahead of major shower or thunderstorm systems, where air is being drawn into the system. The individual cloud elements persist for a while, but decay in the rear of the convective system.

Stratus (St)

Stratus is the lowest cloud genus, and its base is only rarely higher than about 500m. It frequently shrouds the tops of high buildings and is essentially the same as fog, which may be regarded as Stratus cloud at ground level. A layer of Stratus is usually fairly uniform in density, but this may range from being fairly thin (in the variety translucidus) to being so dense (opacus) that it completely hides the position of the Sun and Moon. It may then appear very dark and oppressive. Its base is usually moderately well defined, but is soft and indistinct (unlike the base of Cumulus or Stratocumulus clouds). The base may show undulations (undulatus), but when seen from above, the top of the layer is usually even and featureless.

The cloud particles in Stratus are normally simple water droplets, so that when it is thin, coronae may appear round the Sun or the Moon. Although it sometimes contains tiny ice crystals, halo phenomena (which have been reported) are extremely rare.

The most common method by which Stratus is formed is when warm, moist air is carried over a cold surface, either land or sea. The exact nature of the cloud layer is strongly dependent on the wind speed, and the difference in temperature between the air and the surface. At low wind speeds there is very little turbulence in the slow drift of air, and the lowest layer of air cools first (even with a small temperature difference), which produces a layer of ground fog. At a slightly higher wind speed (3–6m/s or ~10–20kph), turbulence becomes significant and causes the cooling to be spread through a greater depth of air. Stratus then starts to form at the top of the mixed layer and, if the mixing and cooling persist, the cloud will thicken downwards towards the surface. The tops of quite modest hills may be in clear air above the cloud. At higher wind speeds, mixing and cooling take place within a much deeper layer of air and Stratus may not form at all. On occasions, however, when there is a very large temperature difference, Stratus may form very close to the surface. Warm moist air flowing over extremely cold, oceanic water produces the notorious 'Force 10 fog' that is encountered in the Shetlands.

Stratus that forms over the sea may be carried inland. On the eastern coasts of Scotland and northern England, this form of low-lying cloud is known as 'haar'. It tends to form in spring and early summer, when a moderately warm easterly airstream crosses the cold waters of the North Sea. It generally begins as ragged patches of Stratus fractus, but gradually thickens into Stratus nebulosus.

Because of the way in which Stratus forms under relatively still conditions, the tops of hills or even very high buildings may be in clear air and sunshine above the cloud layer. Frequently, however, the wind may be stronger and force humid air to rise above hill and mountain slopes so that it is the higher ground that is shrouded in cloud, while the lower slopes, the valleys and surrounding plains are cloud-free.

Stratus often arises when thin fog over low-lying ground begins to lift in the morning. Sunlight penetrating through the thin fog starts to warm the ground and the layer of fog lifts to become Stratus. The heat gradually warms the lowest layer of air, and the cloud then begins to disperse. It may also disappear if the wind rises, and turbulence mixes a deeper layer of air. Sometimes patches of Stratus fractus linger in valleys and along the sides of mountains. Occasionally these patches of cloud drift up the hill slopes on a gentle flow of air. Such patches are known as 'call-boys' in certain parts of England.

Cross references

Cirrus fibratus p.60–61
Cumulonimbus p.73
Cumulus fractus p.71
Fog p.102–104
Nimbostratus p.67
Stratocumulus p.68

FAR LEFT Stratus opacus cloud enshrouding the top of Snowdon. *(Dave Gavine)*

LEFT This Stratus nebulosus is fairly thin, allowing the Sun's position to be seen, but is otherwise completely featureless. *(Author)*

RIGHT Stratus fractus in the warm section of a depression. *(Author/ Dave Gavine)*

FAR RIGHT Stratus undulatus with an unusual striated appearance. *(Author/ Duncan Waldron)*

Stratus may sometimes arise from a layer of Stratocumulus, with the failure of the weak convection within the layer that otherwise maintains the individual elements and the gaps between them. The base of the Stratocumulus layer becomes indistinct, losing its pattern of lighter and darker areas, and may also lower towards the ground. Conversely, if the wind carries a layer of Stratus over higher ground, or if convection sets in within the layer, Stratus may change into Stratocumulus.

On rare occasions, Stratus (especially Stratus undulatus) may develop a fibrous effect with darker streaks, somewhat like the fibrous form seen in the much higher Cirrus fibratus. Apart from being much darker, however, Stratus cloud is very obviously much closer to the ground.

Stratus often occurs as individual patches of Stratus fractus, before merging into a layer of Stratus nebulosus. Such patches show a superficial resemblance to Cumulus fractus, but the different circumstances under which these forms occur makes it easy to distinguish between them.

A form of ragged Stratus patches, known as pannus, often occur below clouds such as Nimbostratus and Cumulonimbus that are producing moderately heavy precipitation. This is actually an accessory cloud, and will be described later.

Stratus species		
Stratus fractus	(St fra)	Broken Stratus
Stratus nebulosus	(St neb)	Uniform Stratus
Stratus varieties		
Stratus opacus	(St op)	Opaque Stratus
Stratus translucidus	(St tr)	Translucent Stratus
Stratus undulatus	(St un)	Undulating Stratus
Stratus secondary features		
praecipitatio	(pra)	

Cumulus

Cumulus clouds are very familiar. They are heaps of white cloud, usually with fairly distinct outlines and flat, darker bases. They are fair weather clouds, and only one species, Cumulus congestus, sometimes gives rise to any form of precipitation.

Cumulus form from rising thermals, which, as they rise, cool and eventually reach the dew point where the moisture that they carry condenses into cloud droplets. Because this condensation level is at essentially the same altitude over a fairly large extent, all the cloud bases occur at the same level. When there are a number of Cumulus clouds in the sky, it is usually very evident that all the bases are at the same height. This cloud base is normally higher in summer than in winter, because the air is normally drier and warmer, so the thermals need to rise to a higher level for condensation to occur. Similarly, the cloud base is often higher in the afternoon than it is early in the morning.

When Cumulus clouds begin to form, early in the day, the thermals are small and tend to

BELOW RIGHT Well-developed, 'fair-weather' Cumulus clouds. *(Author)*

FAR LEFT Decaying Cumulus fractus behind a passing frontal system. *(Author)*

LEFT Cumulus clouds decaying with the decline in convection at the end of a day, with some suggestions of a layer of velum. *(Author)*

LEFT Clearly defined Cumulus humilis clouds, with Cirrus above, ahead of an advancing warm front. *(Author)*

mix readily with the surrounding air, causing the tiny cloud droplets to evaporate fairly quickly. For this reason clouds forming early in the day are often small, ragged and dissipate quickly. These ragged clouds are known as Cumulus fractus. Later in the day the clouds become larger, more persistent and usually show signs of active growth well above the condensation layer. However, one specific feature of Cumulus clouds is that the individual clouds remain separate from one another, with clear sky between them, even when they increase in size later in the day.

All the time that the thermals persist, Cumulus clouds exhibit rounded tops – an indication that the air within them is still rising and that they are continuing to grow upwards. When the air stops rising, the clouds start to decay as cloud droplets evaporate into the surrounding air. Towards the end of the day, when solar heating declines in the later afternoon, unless other factors are at work Cumulus clouds may decay rapidly, becoming smaller in extent and fewer in number. The exact form will depend on the conditions that have prevailed during the day, but they often become small, ragged tufts of cloud, similar to the form they assumed early in the day. At such a stage they are again known as Cumulus fractus.

When Cumulus grow beyond the fractus stage they develop flat bases and a gently rounded top. If they are much longer than deep, with a distinctly flattened appearance, they are known as the species Cumulus humilis. Such clouds often occur early in the day, but if they continue to grow will turn into Cumulus mediocris. Often, however, Cumulus humilis arise when conditions tend to suppress convection, which limits upward growth and leads to flatter tops to the clouds. This frequently occurs ahead of the warm front of a depression, when it is common to see high-level Cirrus above low-level Cumulus humilis. Similar clouds occur under anticyclonic conditions when the subsiding (and warming) air also prevents the thermals from rising far into the atmosphere.

The next stage in the growth of Cumulus clouds is Cumulus mediocris. Here, the clouds show obvious signs of upward growth, with rounded heads at the top of the individual clouds. The overall depth of the clouds, which may assume a roughly triangular shape, remains less than or approximately the same as the width of the base. Such clouds may begin to show signs of 'leaning' downwind if there is a moderately strong wind blowing.

LEFT Moderately widespread Cumulus mediocris developing by mid-afternoon. *(Author)*

RIGHT An active Cumulus congestus cell, on the point of becoming Cumulonimbus calvus, with lower, less active Cumulus clouds. *(Author)*

FAR RIGHT Contrasting cloud colours. The darkest cloud would appear brilliant white in direct sunlight, but is in the shadow of a nearby large Cumulus cloud, *(Author)*

Even larger Cumulus are known as Cumulus congestus. Here the cloud is obviously growing very vigorously and is generally much deeper than the width of the base. The top is usually brilliantly white, and tends to resemble a cauliflower. The top is sharply outlined, but shows no signs of a fibrous structure, although it may fray out slightly as it decays. These clouds are signs of strong convection, and in the tropics they regularly produce precipitation. At middle latitudes in summer they may be deep enough to produce a shower of rain.

Cumulus congestus clouds are quite easily confused with Cumulonimbus, which do, indeed, develop from them. In winter at middle latitudes, a convective cloud – rather than layer clouds – that produces heavy precipitation is likely to be Cumulonimbus. At other times, if there is any doubt, if there is no precipitation in the form of hail, nor lightning and its associated thunder, it is convention to describe a deep cloud as Cumulus congestus. Being high clouds, Cumulus congestus are also the most likely to be accompanied by the accessory cloud known as pileus ('cap cloud'). If they are producing heavy precipitation, ragged pannus clouds may form beneath them. Because the convection within Cumulus congestus is so strong, they may occasionally produce tuba, *ie* funnel clouds, waterspouts or landspouts.

Cross references

Altocumulus p.64
Cumulonimbus p.73
Pannus p.74
*Pileus p.75**
Rain p.96
Stratocumulus p.68
Waterspouts and landspouts p.119

Cumulus species		
Cumulus fractus	(Cu fra)	Broken Cumulus
Cumulus humilis	(Cu hum)	Flattened Cumulus
Cumulus mediocris	(Cu med)	Medium Cumulus
Cumulus congestus	(Cu con)	Heaped Cumulus
Cumulus varieties		
Cumulus radiatus	(Cu ra)	Cumulus in rows
Cumulus accessory clouds		
pileus	(pil)	
velum	(vel)	
pannus	(pan)	
Cumulus secondary features		
virga	(vir)	
arcus	(arc)	
praecipitatio	(pra)	
tuba	(tub)	

Cumulus clouds are generally described as being white, but like all clouds their colour depends on their exact illumination and on the background on which they are seen. Their tops do appear blindingly white when bathed in bright sunshine, but other, lower regions may appear various shades of grey, or have a bluish tint, in particular when they are thin, ragged and evaporating. When in the shadow of other clouds they may appear almost black, especially when they are seen against the background of more brilliantly illuminated clouds.

Although often created by thermals rising from the surface, Cumulus clouds may also arise through the break-up of a layer of Stratus, Stratocumulus or Altocumulus. A layer of Stratus, for example, that has formed overnight, may start to rise and break up into Stratocumulus when solar heating begins to take effect in the morning, and the Stratocumulus itself may then fragment into individual Cumulus clouds. The existence of Cumulus clouds is always an indication of a certain degree of instability in the atmosphere.

Conversely, Cumulus may evolve into another cloud genus. They may, for example, grow upwards until they reach an inversion, where temperature increases with height. There, they spread out and gradually cover the sky in a layer of Stratocumulus – individual 'pancakes' of cloud with narrow breaks between them. When the inversion is higher, they may similarly form a layer of Altocumulus.

Although Cumulus clouds are water-droplet clouds, they do not generally give rise to rain. An indication of this is the fact that their bases are flat – any rain-clouds usually have ragged bases, from which it is possible to see columns of precipitation descending. One species, Cumulus congestus, does produce showery rain, but this occurs only in summer in temperate regions, although it is a common source of rain in the tropics.

In general, Cumulus clouds are not readily confused with other genera, although small, relatively closely spaced Cumulus seen at a distance may be difficult to differentiate from Stratocumulus and Altocumulus. Closer examination (with binoculars, for example) will normally reveal any signs of upward growth and whether the bases are tending to merge. If not, then the clouds may be taken to be Cumulus. Some difficulty may arise with the species known as Cumulus congestus, because such clouds do share many features with Cumulonimbus.

Clouds of great vertical extent

Cumulonimbus

Cumulonimbus clouds are the giants of the atmosphere. They may tower through all three étages, with bases close to the ground, but with tops that reach the tropopause. Even relatively small Cumulonimbus may produce heavy showers of rain, while large clouds may be the source of lightning (and its accompanying thunder), hail and violent winds. In the very largest systems they may even produce tornadoes.

Cumulonimbus clouds may be so large in extent that it becomes difficult to determine details unless one is at a fair distance that allows a clear view of their highest regions. They are extremely dense clouds, which means that they appear brilliantly white when fully illuminated by the Sun, but that the lower parts, in shadow, appear very dark grey or almost black. Their bases are usually very dark and ragged with heavy precipitation in the form of rain, snow or hail. Sometimes the precipitation may be in the form of virga that do not reach the ground.

Cumulonimbus clouds develop from Cumulus congestus and are initially very difficult to differentiate from that species. If the tops of the rising cells have hard outlines they are generally regarded as a sign that the cloud is a Cumulus congestus. The hard outlines start to soften slightly when ice crystals begin to form in the uppermost regions of the cloud. The cloud is then said to be a Cumulonimbus calvus. (This is slightly paradoxical, because 'calvus' is the Latin for 'bald', which would normally be taken to be hard and smooth, rather than slightly soft.) As large numbers of ice crystals form in the uppermost portion of the cloud it takes on a fibrous, striated appearance, and is then known as Cumulonimbus capillatus. With further development the top of the cloud starts to take on a cirriform appearance, and may produce an immense plume of Cirrus that covers a large part of the sky.

If the convection is very vigorous the cloud may reach the tropopause, where the inversion tends to suppress any further upward growth. (With extremely vigorous upcurrents, they may actually penetrate through the tropopause and create a dome of cloud in the lowermost

LEFT A cluster of at least three active Cumulonimbus cells, with the most distant having developed into an anvil (Cumulonimbus incus). *(Author)*

LEFT A well-developed Cumulonimbus incus cloud, advancing towards the photographer. *(Author)*

RIGHT An active convective cell, just developing a hint of softness at its top, and thus becoming Cumulonimbus calvus. *(Author)*

RIGHT The same convective cell, now showing distinct signs of striations, becoming Cumulonimbus capillatus. *(Author)*

Cross references

Hail p.99
Lightning p.114–115
Mamma p.76
Multicell storms p.115
Rain p.96
Snow p.97
Supercell storms p.116
Tornadoes p.121
Tropopause p.10–12

stratosphere. Such heaps of cloud are known as 'overshooting tops'.) When the cloud reaches the tropopause it spreads out and the Cirrus plume takes on an anvil shape. The variety is then known as Cumulonimbus incus. The greatest extent of the anvil plume stretches downwind, and precedes the Cumulonimbus as it is carried along by the wind. Generally, however, the convection is so vigorous that some of the crystals are even carried upwind, giving a shorter, overhanging shelf of cloud that completes the 'anvil' appearance. It is this rear, overhanging portion of the cloud that often exhibits the extreme, almost globular pouches known as mamma.

During the winter, the glaciation level may be very low, so the resulting Cumulonimbus clouds are often very shallow. The precipitation may still be extremely heavy. This is often the case with the 'lake-effect snow' that falls from clouds that occur when extremely cold air from the Canadian Arctic passes across the relatively warm Great Lakes in North America, acquiring considerable moisture as it does so. This is then deposited as heavy snowfalls near the shores of the lakes, such as over Buffalo, New York.

Because the convection within them is so vigorous, Cumulonimbus clouds tend to draw air in towards them. Although the cloud itself may be moving with the gradient wind, close to its base there is an inflow of air which may be strong enough to give the observer the impression that the cloud is 'approaching against the wind'. The inflow of air may produce the features known as shelf or arch clouds (arcus).

Similarly, the inflow of warm air frequently produces secondary, rising cells, which may appear alongside the main body of the cloud. Frequently these secondary cells themselves develop all the characteristics of a Cumulonimbus, so that a cluster of cells at different stages of development moves downwind. Such 'multicell storms' and the even more violent 'supercell storms' will be discussed in more detail later.

Accessory clouds

There are just three types of accessory cloud, so they are quite easy to recognise.

Pannus (pan)

The form known as pannus (or 'scud') consists of small ragged pieces of cloud beneath Cumulus, Cumulonimbus, Altostratus or Nimbostratus. It may be regarded as an extreme form of fractus, and is most commonly seen beneath the Nimbostratus of a depression

Cumulonimbus species

Cumulonimbus calvus	(Cb cal)	Smooth ('bald') Cumulonimbus
Cumulonimbus capillatus	(Cb cap)	Fibrous Cumulonimbus

There are no Cumulonimbus varieties, but all three accessory clouds and the six secondary features often occur with the same cloud or cluster of clouds.

Cumulonimbus accessory clouds

pannus	(pan)
pileus	(pil)
velum	(vel)

Cumulonimbus secondary features

praecipitatio	(pra)
virga	(vir)
incus	(inc)
mamma	(mam)
arcus	(arc)
tuba	(tub)

or Cumulonimbus. It arises when precipitation from the main cloud cools the air beneath it to such a degree that the air reaches its dew point, giving rise to the ragged shreds of cloud. It is not often seen beneath Altocumulus, and only occasionally beneath precipitating Cumulus (Cumulus congestus).

Pileus (pil)
Pileus, or 'cap cloud', is a small cloud hood that is formed when a rising Cumulus or Cumulonimbus thermal encounters a previously invisible humid layer and pushes it upwards into cooler surroundings, causing it to reach the dew point and become visible as a 'cap' draped over the rising tower. Initially the pileus may be quite distinct and separate from the main cloud, but generally the rising thermal penetrates the centre of the cap, which may then appear as a 'brim' around the tower, somewhat like a form of velum. Eventually the flow of air around the thermal incorporates the humid layer into the rising circulation, causing the pileus to disappear.

Velum (vel)
Velum is, like pileus, associated with Cumulus or Cumulonimbus cloud, but in this case consists of a widespread thin sheet or veil of cloud, through which the individual cumuliform towers penetrate. It may be regarded as a thin sheet of Stratus, which was either pre-existing or formed through general uplift as generalised convection set in, before specific, distinct thermals became strong enough to produce rising Cumulus towers. The cloud form is comparatively rare, but may be extremely persistent, remaining in the sky after convection has died down late in the day and the cumuliform clouds have dispersed.

FAR LEFT Extensive ragged pannus developed in the humid air beneath a layer of Nimbostratus. *(Author)*

LEFT Pileus: a layer of humid air, lifted to its dew point by a rising convective cell. *(Author)*

LEFT A layer of velum, remaining behind following the passage of a cluster of Cumulonimbus and Cumulus convective cells. *(Author)*

Supplementary features

Although clouds may assume a vast range of shapes, there are a few (six) distinctive forms which have been given specific names, and which are relatively easy to recognise.

Arcus (arc)
Arcus is a long, dense roll of cloud, sometimes with ragged edges, that occurs along the lower leading edge of the main cloud, which is normally Cumulonimbus, but may occasionally be a large, active Cumulus congestus. There is normally a fairly distinct clear layer, of fairly uniform depth, beneath the roll of cloud, producing the appearance of an arch across the sky. Depending on conditions, arcus may appear light in colour or dark and threatening. The term 'roll cloud' is often generally applied to arcus, but that term, and 'shelf cloud', are better applied to specific

Cross references
Cumulonimbus p.73
Cumulus congestus p.72–73
Morning Glory p.106
Multicell storms p.115
Supercell storms p.116

LEFT A shelf cloud (arcus) ahead of a major Cumulonimbus cluster. Air is rising along the upper, smooth surface into the advancing cloud. *(Author)*

RIGHT **Cumulonimbus incus, part of an advancing cold front, visible through the unusually clear skies ahead of the front.** *(Author)*

RIGHT **The overshooting tops, where vigorous Cumulonimbus cells have penetrated the tropopause and into the lower stratosphere, photographed from orbit.** *(NASA)*

forms of arcus with very distinct features. Arcus is particularly prominent in major multicell and supercell storms. Probably the most dramatic form of roll cloud is that associated with the squall line observed in northern Australia and known as the Morning Glory.

Incus (inc)

This supplementary feature, the 'anvil cloud', is hard to confuse with any other. It occurs solely with Cumulonimbus clouds (Cumulonimbus incus), and forms when the rising convective cells reach an inversion, but possess insufficient energy to penetrate it. The cells are forced to spread out sideways, giving the characteristic flattened shape. Although inversions may occur at any altitude, the major one that separates the troposphere from the stratosphere (the tropopause) arrests the development of even the most vigorous Cumulonimbus clouds.

The low temperatures cause glaciation to take place and the resulting ice crystals are borne away downwind by the generally higher wind speeds at that altitude. Because of the vigorous convective overturning in the rising cells, the cloud normally also spreads out on the upwind side, although to a lesser extent. The small extension upwind and the larger one downwind produce the characteristic anvil shape. The main cloud moves downwind, so the extension on the upwind side actually trails behind the main body of the Cumulonimbus. Another supplementary feature, known as mamma, often forms beneath the trailing edge of overhanging cloud.

Although the tropopause arrests the upward growth of Cumulonimbus cells, the most vigorous cells may penetrate a short distance into the stratosphere, giving rise to what are known as 'overshooting tops', small heaps of cloud above the main flattened layer. Such domes of cloud may sometimes be seen from the ground, but are often obscured by the lower clouds. They are often clearly visible from aircraft or in satellite images.

Mamma (mam)

Bulges or pouches on the underside of a cloud are known as mamma. (The word means 'breast' or 'udder'.) They may be found beneath a wide variety of cloud types (Ci, Cc, Ac, As, Sc and Cb), although the most striking are those that arise beneath the overhanging anvils of Cumulonimbus incus. There is no general agreement as to the exact mechanism by which they are produced and it would seem that the actual means might differ in different cloud genera. They are produced when downdraughts drag cold air down into a lower, warmer, humid layer, causing it to cool to its dew point, creating cloud droplets. The effect has sometimes been described as 'upside-down convection' because it mirrors the circulation within rising thermals.

The mamma beneath most of the cloud types are relatively small and inconspicuous, but there is often a major exception in the case of mamma beneath Cumulonimbus anvils. There they may be large, rounded, almost globular pouches, which may appear extremely dramatic when they are lit by low sunlight, especially when seen against a dark background. The mamma are particularly large because it seems that the upper side of the anvil readily radiates heat away to space, producing a very significant drop in the air temperature. Such mamma may sometimes persist even after the Cirrus anvil itself has dispersed. Mamma may also arise beneath the main body of a Cumulonimbus cloud where there are suitable downdraughts, and these,

Cross references

Cumulonimbus p.73
Glaciation p.53
Multicell storms p.115
Supercell storms p.116
Tropopause p.10–12

rather than resembling pouches, often take the form of long, distorted tubes, which have been likened to the trunks of elephants.

Praecipitatio (pra)

The general term for the visible streaks of precipitation that are observed to reach the ground is praecipitatio. The term applies to a whole range of precipitation: from a light drizzle to torrential rain, from a light dusting of snow to highly destructive hail. The important factor is that the individual particles do survive to reach the ground. If snow crystals or ice pellets melt and then evaporate, or raindrops evaporate before they hit the surface – even if at very low altitude – the trails of precipitation are known as another supplementary feature and by another name: virga. Praecipitatio is normally closely associated with Cumulus congestus, Nimbostratus and Cumulonimbus but may, on occasion, accompany Altostratus, Stratocumulus and even Stratus, although in the last two cases in particular the precipitation may be extremely light and in the form of drizzle or, if the temperature is low enough, of snow grains.

Tuba (tub)

The term tuba is applied to any approximately cylindrical or tapering column of cloud that extends downwards from the main cloud base towards the surface. It is frequently commonly described as a 'funnel cloud'. The column may be rotating or non-rotating, but the term carries no implications for the method by which the column of cloud has been formed. It may be applied to small, innocuous funnels hanging beneath the main cloud (usually vigorous Cumulus congestus or Cumulonimbus) or to incipient tornadoes. It is, in any case, completely distinct from the various devils and whirls that extend upwards from the ground.

If a tuba does extend down to the ground, it will then become a specific phenomenon and be known by a completely different name. Depending on its exact method of formation it may be a landspout or waterspout (associated with very vigorous convection), or a true tornado, which arises through a complex process found in supercell storms.

Funnel clouds (tuba) are not particularly rare. Apart from strong convective clouds they are also found on gust fronts and most major storm systems (including multicell storms, supercell storms and tropical cyclones).

Virga (vir)

Virga are trails of solid or liquid precipitation that may be seen falling from clouds, but that evaporate before they reach the ground

Globular mamma below part of the overhanging anvil behind a complex, major Cumulonimbus cluster. *(Author)*

LEFT Distorted mamma below a large, individual Cumulonimbus cloud. *(Author)*

Cross references

Cumulonimbus p.73
Cumulus congestus p.72–73
*Drizzle p.*95*
Hail p.99
Nimbostratus p.67
Rain p.96
Snow p.97
Snow grains p.98
Landspouts p.119
Multicell storms p.115
Supercell storms p.116
Tornadoes p.121
Tropical cyclones p.124
Waterspouts p.119
Whirls and devils p.119

LEFT Praecipitatio (rain) beginning to fall from a single shower (Cumulonimbus) cloud. *(Author)*

LEFT A very typical funnel cloud (tuba), suggesting the existence of convection buried within the layer of relatively featureless stratiform cloud. *(Author)*

Cross references

Fallstreak holes p.94
Wind shear p.166

(unlike praecipitatio). They are also known as 'fallstreaks', and occur with a wide range of cloud types: Cirrocumulus, Altocumulus, Altostratus, Nimbostratus, Stratocumulus, Cumulus and Cumulonimbus. Indeed, Cirrus, particularly Cirrus uncinus, might be considered to consist solely of virga, because there is usually a generating head – even though it may be small and inconspicuous – where the ice particles originate before falling out, and then trail behind in the lower winds, and either sublime directly into water vapour or melt and then evaporate.

The appearance of virga strongly depends on the properties of the layer or layers into which they fall. Sometimes (as in Cirrus) they may be essentially horizontal, and at other times, when they descend into a deep layer that is moving at a constant speed, they may appear almost vertical. Frequently they may display a distinct change of direction partway down. This may occur either when there is a sudden wind shear between layers, or else when ice particles that have been falling vertically reach a warmer layer and melt into water droplets, which then trail behind in the lower, slower-moving layer. Under very unusual circumstances, when a lower layer of air has been moving considerably faster than one above, virga have actually been seen ahead of the cloud in which they have originated.

Virga are often associated with what are known as 'fallstreak holes' or 'hole-punch clouds', where a distinct hole – often strikingly circular in shape – appears in a layer of Altocumulus or Cirrocumulus. These will be described in more detail later.

Nacreous clouds and ozone holes

There are some striking atmospheric phenomena that are not closely related to weather conditions at the surface, but which are always worth observing and recording. The most conspicuous of these are nacreous and noctilucent clouds and auroral displays.

It is hard to believe that a beautiful type of cloud is closely involved in the creation of the ozone holes, one of the major effects that human activity has had on the atmosphere. But this is the case with nacreous clouds, which, although quite rare, are so striking that they are often reported widely in the media whenever a major display takes place.

They are commonly known as 'mother-of-pearl clouds' from their iridescent colours like those seen inside oyster and ormer shells. As with most atmospheric optical phenomena, their appearance may vary from exceptionally striking

RIGHT Dense virga beneath Altocumulus floccus clouds. *(Author)*

FAR RIGHT These long fallstreaks were evaporating in the air very close to the ground, but would therefore be classed as virga rather than praecipitatio. *(Author)*

RIGHT Virga below an Altocumulus floccus element. Ice crystals were falling vertically, then melting into water droplets that trailed behind the generating head. *(Author)*

pure bands of colour to a more subdued, limited range of tints. At sunset, for example, the clouds may initially show a whole range of colours, which are often delicate pastel shades. As the Sun sinks lower below the horizon, the colours shift towards orange and red (because the shorter wavelengths are absorbed during the longer path through the atmosphere) until finally the clouds appear red. The opposite sequence occurs at sunrise, of course. Like the iridescence seen in other clouds, the bands of colour tend to run parallel to the edges of the clouds, where the particles are of similar sizes.

The more technical name of 'polar stratospheric clouds' (PSCs) gives an indication of the nature of nacreous clouds. They consist of ice crystals and occur in the lower stratosphere, from just above the tropopause (and thus at altitudes of about 15km) up to approximately 30km. They are seen just after sunset or before sunrise, when the Sun is below the horizon and observers on the ground are in twilight. The clouds themselves are high enough to be illuminated by direct sunlight. Only at reasonably high latitudes are conditions right for this to occur, but PSCs form in both the Arctic and Antarctic regions. They are more often noticed after sunset than before dawn, but this is purely because more people are around in the evenings than in the early morning. The clouds may be so striking in appearance that major displays are widely reported and often come to be mentioned on radio and television news programmes.

There are actually three types of polar stratospheric clouds and these types are related to the structure of the cloud particles, which may be moderately complex. The low temperatures (about -78°C) at the observed heights cause a substance known as nitric acid trihydrate ($HNO_3.3H_2O$) to be deposited on sulphuric acid nuclei. These initial particles are invisible to the human eye but may form multiple layers that extend over vast distances amounting to thousands of kilometres.

Although the air at such altitudes is very dry, if the temperature drops even lower (below -83°C), the second type of PSCs may form. Water ice freezes on to the nitric acid trihydrate particles. When the uplift is rapid, as it tends to be in lee waves, many tiny crystals (of different sizes) form, producing the beautiful colours through diffraction effects. These are the most common type of nacreous clouds in the northern hemisphere.

When the air cools very gradually, such as at the onset of winter, the water ice accumulates very slowly and tends to create much larger particles. These appear colourless, so the clouds are white, with perhaps just a touch of iridescence. This third type of PSC is more common around Antarctica than at northern latitudes.

Nacreous clouds are of interest to professional meteorologists because of their dependence on the exact conditions prevailing in the stratosphere. To amateurs, they are striking events, worthy of observation and photography.

ABOVE Part of a widespread display of nacreous clouds, photographed from North Lincolnshire. *(Peter Roworth)*

Cross references

Iridescence p.134
Wave clouds p.84–87

LEFT Not a nacreous cloud, but the result of a missile launch from Vandenberg Air Force Base in California, and where water vapour from the exhaust has frozen into ice. The distortions in the trail are an indication of differing wind directions at various heights. *(Stephen Pitt)*

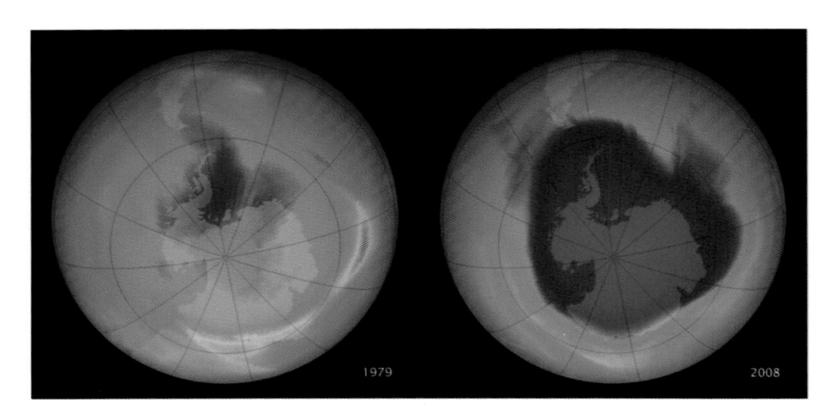

ABOVE A comparison of the sizes of Antarctic ozone holes for 1979 and 2008, where the dark blue and violet areas indicate where there has been significant ozone depletion. *(NOAA)*

Ozone holes

As mentioned earlier, the action of ultraviolet light from the Sun breaks down normal oxygen molecules (O_2), which recombine into ozone (O_3), creating a layer of ozone in the stratosphere. The first description of ozone depletion over Antarctica was published by G.M.B. Dobson, a British scientist, in 1956, but at that time atmospheric conditions in the region were essentially unknown. In 1985, scientists from the British Antarctic Survey discovered that there was an extreme depletion of ozone in a large area above Antarctica, giving rise to what became known as the 'ozone hole'. Because ozone is such an effective shield against a particularly damaging form of ultraviolet radiation, called UV-B, which is known to cause skin cancers and genetic damage in most living organisms, this was of immediate concern.

It was soon established that the ozone depletion was increasing and was being caused by the presence of man-made chemicals in the atmosphere, most notably chlorine (primarily derived from compounds known as chlorofluorocarbons, CFCs), but also bromine and other chemicals. Although their use as aerosol propellants was most commonly mentioned, the chemicals were employed in a wide range of products and processes, and had extremely long atmospheric lifetimes, measured in decades.

Luckily, the international community rapidly took note of the scientific findings – unlike the reaction of some nations to the evidence for climate change and global warming – and an international agreement (the Montreal Protocol) was adopted on 16 September 1987, banning the use of the harmful chemicals. (16 September has since been designated 'World Ozone Day' by the United Nations.) The concentration in the atmosphere of the banned chemicals has been slowly declining, although there are concerns that some of the substances introduced as substitutes may themselves have harmful effects. There are indications that ozone concentrations have now stabilised, although it will be some years before the extent of the depletion may be expected to decrease and the holes shrink in extent.

All three forms of polar stratospheric clouds play a key role in the formation of ozone holes. Although the ozone hole over Antarctica was the first to be discovered, ozone depletion does sometimes occur over the Arctic, but here it is not as extreme. The lack of land masses at high southern latitudes (except for the very tip of South America and the Antarctic Peninsula) means that the wind flow around the world is largely unimpeded, unlike in the north. During the winter, a polar vortex of high-speed winds develops, which essentially isolates the southern polar region from the rest of the atmosphere. Conditions in the south are thus more favourable for the formation of extensive polar stratospheric clouds.

In the north, the land masses and mountains tend to disrupt the flow and mix the air both horizontally and vertically. Although a polar vortex does sometimes form, it tends to be weaker, smaller in diameter and less stable. There is ozone depletion, but no true ozone 'hole'. In recent years the greatest depletion in the northern hemisphere occurred in March 2011. Prior to that, the greatest effect was in 1997.

The colder the winter, the more intense any ozone depletion is likely to be (in both

BELOW Changes in the total area of the Antarctic ozone hole for three years (1998, 2008 and 2010) together with the minimum, average, and maximum areas for the ten years 2000—2009. *(Ian Moores)*

2010 Southern Hemisphere Ozone Hole Area
Ozone Hole Area
30
27
24
21
18
15
12
9
6
3
0
August
September
October
November
Dec
2010
2008
1998
00-09 Mean
00-09 Max
00-09 Min

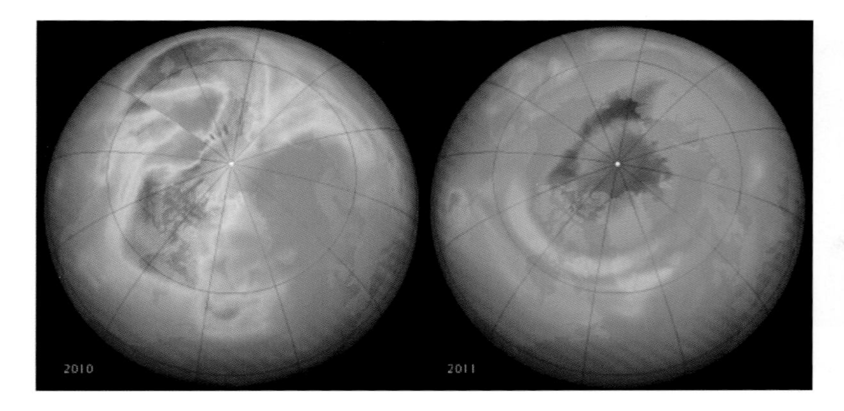

ABOVE **Northern-hemisphere ozone depletion in 2010 (left) and 2011 (right), as shown by the Ozone Measuring Instrument on NASA's Aura satellite.** *(NASA)*

ABOVE **Nacreous clouds observed in 2011 at the time of the maximum ozone depletion in the northern hemisphere.** *(Per-Andre Hoffman)*

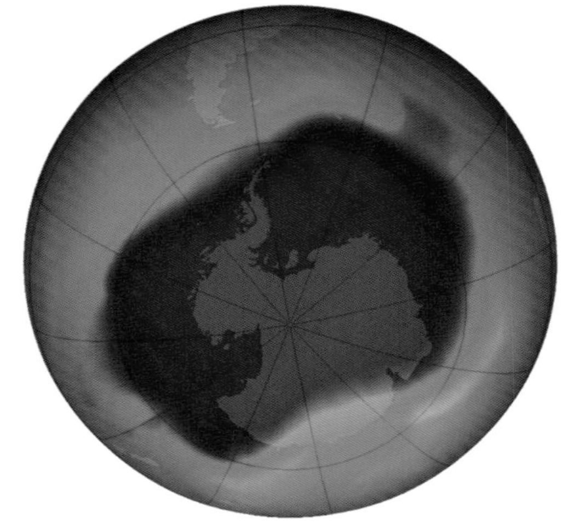

RIGHT **The largest Antarctic ozone hole ever observed, in September 2006. This is thought to have been the result of extremely low temperatures – certainly the lowest ever observed – over the Antarctic that year.** *(NASA)*

hemispheres). In a very cold winter, a lesser ozone-depleted area may also form over Tibet and the immediately adjacent regions, such as the Hindu Kush, but this is far less intense than the amount of reduction found over the Arctic and, particularly, over the Antarctic. When winter conditions in the north are exceptionally warm the temperatures may be too high for PSCs to form and ozone depletion is correspondingly less.

When polar stratospheric clouds are present during the long polar night, various chlorine compounds accumulate on the surface of the cloud particles. Throughout the long period of darkness there is no light to drive the photochemical reactions that destroy ozone. With the return of sunlight in spring, however, not only can photochemical reactions take place, but warming also destroys the cloud particles, releasing the active compounds into the atmosphere. The greatest ozone depletion therefore occurs in spring, when the effects of the Antarctic ozone hole may extend well towards the north, reaching the southernmost areas of Australia, New Zealand and South America. Although the most obvious effect on humans might be an increase in skin cancers, long-term genetic effects on other organisms might be even more significant. The waters of the Southern Ocean around Antarctica are some of the most productive in the world, and it is feared that increased exposure to UV-B radiation may cause major damage to the oceanic plankton on which the whole food chain ultimately depends.

An ozone hole disappears when the polar vortex breaks down, and air from lower latitudes brings ozone into the polar region.

Cross references

Iridescence p.134
Polar vortex p.80–81
Stratospheric ozone layer p.12

BELOW **The variation in Antarctic ozone throughout the year. The decrease when the southern spring arrives in August and September is extremely rapid and dramatic.** *(NOAA/NASA)*

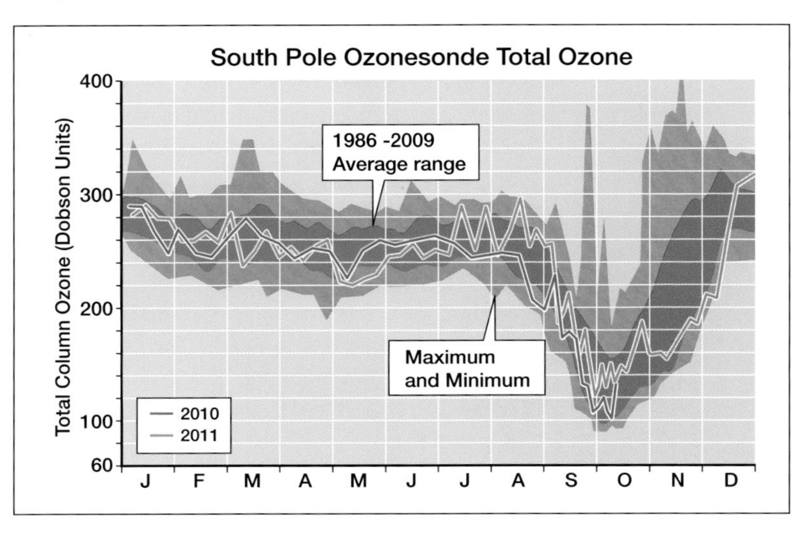

RIGHT A bright display of noctilucent clouds, photographed from Finland. *(Peka Parviainen)*

Noctilucent clouds

There is a rare form of cloud that is visible only occasionally in the summer, and if the observer is at a high enough latitude. These are noctilucent ('night-shining') clouds (NLC), and they appear in the direction of the pole during the middle of the night. Like nacreous clouds they are illuminated by the Sun when it is well below the horizon, but the observer is in darkness. In this case, however, the clouds are high, extremely high, occurring within the mesosphere at altitudes of some 80–85km, far above nacreous-cloud heights (15–30km). They are the highest clouds ever observed, and as such pose problems for scientists attempting to explain their occurrence. The clouds have a short observing season of about six weeks before and after the summer solstice (June and December in the northern and southern hemispheres, respectively).

In appearance, noctilucent clouds are like wispy Cirrus or Cirrostratus, and they have often been mistaken for those much lower cloud types, which may sometimes be seen as dark silhouettes against them. They generally have a distinctive bluish-white colour, although they may appear slightly yellowish in tint both early and late in an individual display. The cloud particles exist as a very tenuous layer that is normally so thin that it is often completely invisible when it is directly overhead. The apparent structures actually arise because the layer undulates under the influence of high-altitude atmospheric waves, some of which – if not all – have been generated as mountain waves at the surface far beneath and have propagated upwards. The greater the length of the observer's line of sight through the undulating layer, the denser the cloud structure appears to be.

The clouds are known to consist of ice particles and are believed to form at the very coldest part of the atmosphere, the mesopause, and to grow and become visible as they drift down to their observed height. But there are two major unresolved problems. Where does the water come from that is deposited as ice, and what are the freezing nuclei? It is difficult to imagine any method by which water could be transported from lower levels (primarily from the troposphere) to such great altitudes. The stratospheric temperature inversion effectively blocks any convective form of transport, and although it was once considered that violent volcanic eruptions might eject water to such heights, the energy involved would be so extreme that the process is nowadays considered to be incapable of projecting water vapour to heights greater than the lower stratosphere. As we saw earlier, ice crystals form on freezing nuclei, and similar considerations apply to these. They were once considered to be ash particles or sulphur dioxide of volcanic origin, but even the most energetic eruptions are unlikely to be able to transport any material to heights of 80–100km.

Certainly there is now an unconfirmed suspicion that the water enters the atmosphere from outside, *ie* that it originates from interplanetary space, and is introduced as particles of ice from comets or in the form of micrometeorites. If the water is in the form of ice particles (which would readily grow to larger sizes in the presence of water vapour), these would have formed with interplanetary dust particles as nuclei. If the water is in the form of vapour, we still require some form of freezing nuclei, and these might be interplanetary dust particles or possibly ion clusters that have been created by energetic incoming cosmic rays. At present the correct explanation remains unknown.

Noctilucent-cloud displays seem to be becoming more frequent, although it is difficult to know whether this is because there are increasing numbers of observers who are aware of their existence (and report them) or whether their frequency is truly increasing. There are some suggestions that displays are linked to the increasing frequency of rocket launches, which certainly inject large quantities of water vapour into the upper atmosphere. For these and many other reasons, NLC are of considerable interest to scientists studying the

Cross references

Cirrostratus p.63
Cirrus p.60
Glaciation p.53
Mesosphere p.13
Atmospheric layers p.10
Composition of the atmosphere p.14

upper atmosphere, and any observations are of potential significance. Many reports are received from amateurs who normally observe aurorae, and they often form part of the programmes of auroral observing groups such as that of the British Astronomical Association. Because the NLC season occurs during high summer, when auroral observation is difficult, or impossible, and the techniques are similar, most observers contribute information about both types of event.

The formation of clouds

There are two dominant factors governing the nature of the clouds that form at any one time. These are:

- The growth of condensation droplets to form the cloud.
- The stability of the atmosphere.

Cloud droplets form when air cools below its dew point. This may be caused by two basic reasons:

- The air is forced to ascend and cool in doing so.
- The air comes in contact with a cold surface.

Under certain uncommon circumstances cloud may also result from the mixing of two air masses, close to saturation, with differing temperatures.

There are various ways in which air may be impelled to rise, when it will cool at the dry adiabatic lapse rate (DALR) until condensation occurs, and then at the saturated rate (SALR). There are three main mechanisms of uplift:

- Through lifting by being undercut by colder air (frontal lifting).
- Through forced ascent over hills or mountains (orographic ascent).
- Through convection, normally through heating from below (convective heating).

Frontal lifting

The stability of the atmosphere determines the nature of the clouds that are formed. With a stable atmosphere, the air must be forced to ascend for adiabatic cooling to take place,

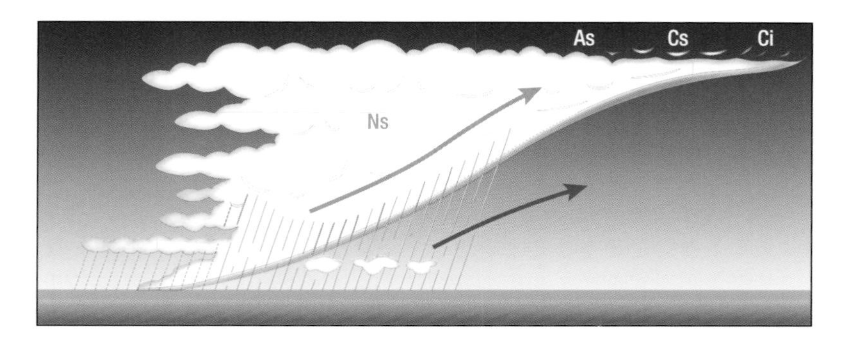

ABOVE The gentle slope (1:100 or 1:150) of an ana warm front (here much exaggerated) means that any rainfall is greatly prolonged. *(Dominic Stickland)*

either through frontal lifting or orographic ascent. At a warm front, particularly an ana warm front, the ascent is gradual and smooth. It takes place over a wide area and the resulting cloud is a basically stratiform sequence: Cirrus, Cirrostratus, Altostratus and Nimbostratus. Only very occasionally does any convective activity occur, mixed in with the predominant stratiform clouds. When the warm air is subsiding, as at a kata warm front, then the succession of clouds is absent, with the Cumulus ahead of the front gradually increasing into Stratocumulus, which becomes very thick around the frontal zone.

At a cold front, where the warm air is being undercut and lifted by the advancing cold air (at an ana cold front), the sequence of stratiform clouds is reversed, with the low Nimbostratus giving way to Altostratus and Cirrostratus with Cirrus trailing behind. Often, however, because the cold air is advancing over a relatively warm surface, there is some convective activity, with Cumulus congestus and even Cumulonimbus. A typical cold front will show this mix of cloud types, and is sometimes described as a 'passive' cold front. Sometimes, however, the transition from warm air to cold air is extremely abrupt and the frontal zone is much closer

Cross references

Frontal systems p.30
Lapse rates p.51
Stability and instability p.50

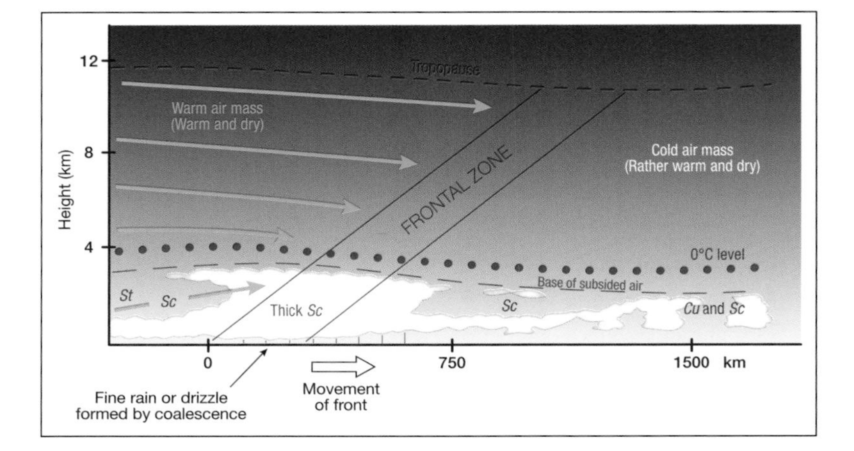

BELOW At a kata warm front, all the clouds are subdued, and any precipitation will be light, or even non-existent. *(Dominic Stickland)*

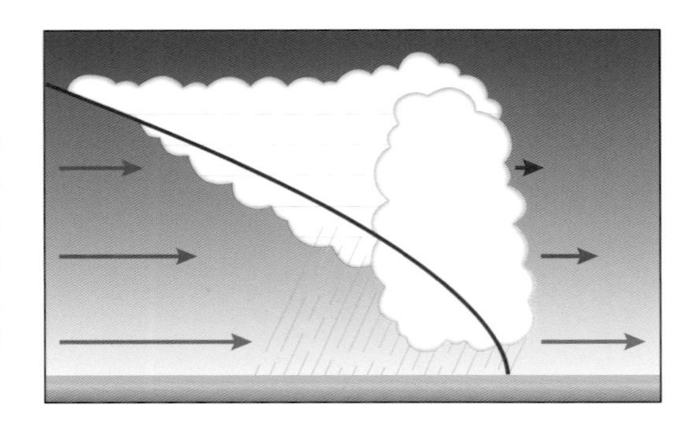

RIGHT The slope (1:50 to 1:75) of an ana cold front means that the precipitation is shorter-lived than at the corresponding warm front. *(Dominic Stickland)*

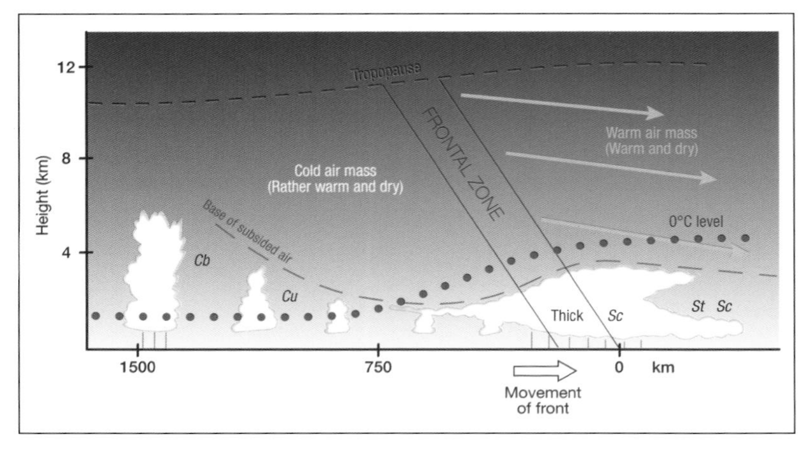

ABOVE At a kata cold front, the thick stratocumulus and any accompanying slight precipitation gradually give way to cumuliform cloud in the cold air. *(Dominic Stickland)*

to vertical. Such an 'active' front is less often associated with a depression, but occurs in the form of an active squall line of Cumulonimbus clouds, frequently accompanied by major thunderstorms.

By contrast, at the kata cold front of a depression (where the warm air is descending) the Stratus and Stratocumulus of the warm sector becomes thick Stratocumulus at the frontal zone, and gradually gives way to convective cloud behind the front.

Orographic lifting

With orographic lifting, the ensuing cloud and its formation and decay will greatly depend on the air's stability. With stable air that is forced to rise over hills or mountains, the result will depend upon the exact condensation level. If it is low, the tops of the high ground will be shrouded in fog (actually Stratus at ground level), but if there is no precipitation the cloud will clear to leeward at approximately the same level when it descends and undergoes adiabatic warming. (When precipitation falls out of the cloud over the high ground, the rise in temperature on the leeward side may be very considerable in what are known as föhn conditions, to be described later.)

When the condensation level is above the peaks, wave clouds are likely to be formed. A single wave cloud (lenticular Stratocumulus, Altocumulus or Cirrocumulus) may form above the top of the hill or mountain, separated from it by clear air. Again, such a 'cap cloud' will disperse as the air descends after the obstacle. Depending on the exact conditions a whole train of wave clouds may appear downwind in the crests of the otherwise invisible waves created by the high ground. Sometimes, a more or less closed, vertical circulation of air, known as a rotor, may form downwind of the barrier. Under the rotor the surface wind blows towards the high ground, and the rotor may be deep enough for a lenticular cloud to form at the top. (This is the case with the 'helm bar' that forms when the helm wind is blowing over Cross Fell in Cumbria.)

Occasionally, there may be no gaps between the individual clouds in the wave-crests, so that there is a long trail of cloud, downwind of the peak. If there is a series of humid layers, one above the other, sets of lenticular clouds may appear, stacked one above the other like a series of inverted saucers. This form is known as a 'pile d'assiettes' (French for 'a pile of plates').

Billows

The formation of billows occurs in a somewhat similar fashion. Here, however, rather than there being a physical obstacle that creates the waves, the wave-train is initiated because there

RIGHT Air rising over high ground will, if it reaches the condensation level, shroud the heights in cloud, which will disperse as the air descends on the leeward side. *(Ian Moores)*

RIGHT Orographic Stratus clouds shrouding the mountains of South Georgia. *(Gordon Pedgrift)*

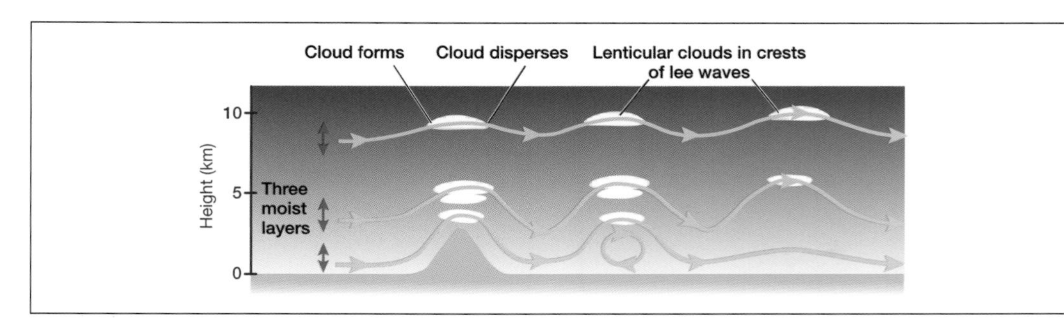

LEFT Humid layers of air will give rise to a succession of lenticular clouds in the crests of lee waves. Under certain circumstances, a lee rotor may form, with the surface wind blowing towards the high ground. *(Ian Moores)*

ABOVE Altocumulus lenticularis on two different levels, generated by a wind from the left. Lower Cumulus clouds are visible, together with the suggestion of a second wave-crest at the higher level. *(Wood)*

ABOVE CENTRE Well-developed wave clouds (Altocumulus lenticularis) above the mountains of Norway, photographed from a meteorological research aircraft. *(Wood)*

LEFT Multiple layers of lenticular (wave) clouds above Boambee, New South Wales. *(Duncan Waldron)*

is vertical wind shear between two layers (the upper layer is moving faster than the lower). This creates rolls of cloud that lie approximately at right angles to the wind direction. Because the boundaries of jet streams are regions of extreme wind shear, they often exhibit well-defined billows.

Although billows exhibit clear sky between the individual rolls, wind shear may also create undulations on the top of a layer of stratiform cloud such as Stratus, Altostratus and Cirrostratus. Usually the base of the cloud layer (the condensation level) is relatively flat but there may be fairly regular undulations in thickness, with thicker cloud in the crests and thinner in the troughs.

On rare occasions this form of wind shear may create Kelvin-Helmholtz waves in the cloud layer. These appear like the breaking waves seen on the sea, but this is an illusion, and they do not 'break'. Such K-H waves are usually extremely short-lived, lasting no more than a couple of

Cross reference

Föhn conditions p.108

BELOW Billows (the undulatus variety) arise in thin stratiform cloud when there is vertical wind shear between two adjacent layers. *(Ian Moores)*

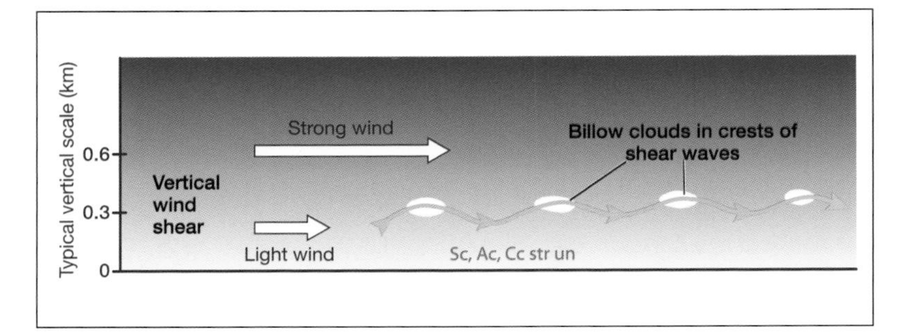

Cross references

Dew point p.164
Jet-stream clouds p.94
Mist and fog p.102–104
Pannus p.74
Radiation fog p.102
Föhn conditions p.108
Sea breezes p.105
Wind shear p.166

ABOVE Marked billows in Altocumulus cloud. Although slightly curved, the billows lay approximately at right-angles to the wind at height. *(Author)*

ABOVE This photograph illustrates how wind shear creates distinct Altocumulus undulatus billows and also wave-like thicker and thinner undulations in a layer of cloud. *(Author)*

ABOVE This train of Kelvin-Helmholtz wave clouds above San Francisco Bay was visible for just a short period of time. *(Peggy Duly)*

ABOVE A large Cumulus congestus cloud that is just turning into Cumulonimbus calvus and which has arisen through uplift over a mountain peak.

RIGHT The banner cloud behind the Matterhorn is best developed (as here) when there is a north-west wind. *(Wikipedia Commons image)*

minutes at the very most, so one needs to react immediately to capture them in photographs.

Billows may sometimes have a superficial resemblance to wave clouds, but there is one vital difference: wave clouds remain stationary in the sky while the wind strength and direction remain constant. Billows, however, move downwind with the general airflow at their height.

Occasionally fine undulations occur that lie parallel to the wind direction. These are known as corrugations and may arise superimposed on larger, transverse billows. They are also found on wave clouds and are often fairly prominent along the trailing edge where such a cloud is dissipating, appearing as narrow 'fingers' of slightly more persistent cloud, pointing downwind.

Instability and orographic lifting

If the air is close to its dew point, orographic lifting and the consequent condensation may release enough latent heat for the air to continue to rise, resulting in convective cloud building up over the high ground. In extreme cases this may lead to major Cumulonimbus clouds that tend to remain stationary over the peaks, sometimes for many hours. The intense precipitation that they may release is often the cause of flash flooding in the neighbouring valleys. A sea breeze that encounters a line of hills some distance inland often produces a line of convective clouds. When sea breezes from both sides of a peninsula meet over the higher ground in its centre, the convective cloud that arises often produces torrential rain and flooding.

A particular form of orographic cloud is the banner cloud, where a plume of cloud streams from the leeward side of a mountain. Famous examples are those exhibited by the Matterhorn and by the Rock of Gibraltar, but they occur with many other peaks, particularly those where the downwind side has a sheer drop, rather than gradually sloping. Isolated peaks seem to exhibit banner clouds more frequently than those that are amongst other high peaks, probably because in the latter case there is more turbulence and shifts in wind direction.

Convective clouds

All cumuliform clouds arise through convection, which is also the mechanism by which some stratiform clouds break up into the perlucidus variety. The clouds of the Cumulus and Cumulonimbus genera are created by thermals that rise from the warm surface, particularly from land that has been heated by the Sun. A film of heated air forms on the surface, and eventually it breaks away into rising 'bubbles' of air, rather like the shape of a hot-air balloon. There is actually a circulation of air: air is rising in the centre, flowing out in a cap at the top and descending around the outside. The descending air tends to be drawn into a wake beneath the bubble of air, mixing surrounding air into the rising air, gradually causing it to erode. Soaring birds and glider pilots fly in circles to remain within the flow of rising air. This type of circulation continues when the air reaches its dew point, with cloud droplets forming in the centre and tending to evaporate in the descending air at the sides.

These thermals are small and only rise about twice their diameter before decaying. Larger thermals form through the combination of the wakes of several adjacent bubbles. The higher in the atmosphere, the larger the rising bubbles, but the lower the temperature difference between the thermal and its surroundings. The rising air will cool at the dry adiabatic lapse rate until it reaches saturation. The initial clouds that form are ragged Cumulus fractus, but gradually the small bubbles combine to produce larger Cumulus clouds. One important factor is the environmental stability above the level at which condensation occurs. If the environmental lapse rate (ELR) exceeds the saturated lapse rate (SALR), even by a small amount, the cloud will continue to grow until it reaches a stable level. With slight instability the clouds will be shallow Cumulus, but with greater instability they will develop further, becoming Cumulus mediocris, Cumulus congestus, or even Cumulonimbus.

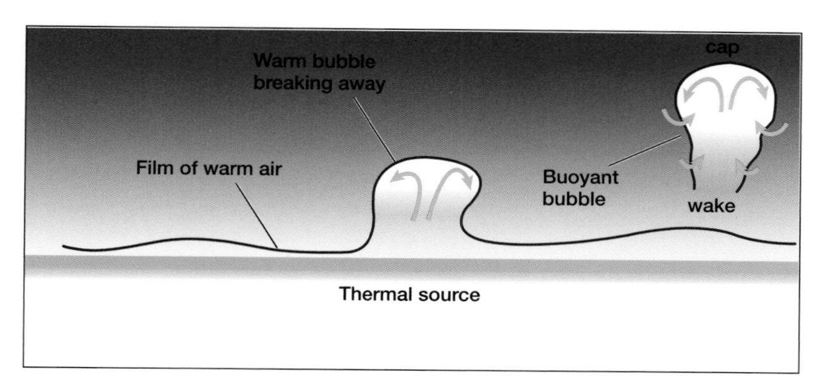

ABOVE Convective bubbles of air occur when a film of warm air over a warm source (usually the ground) breaks free and begins to rise in the atmosphere. *(Ian Moores)*

Cross references

Cumulonimbus p.73
Glaciation p.53
Precipitation p.95

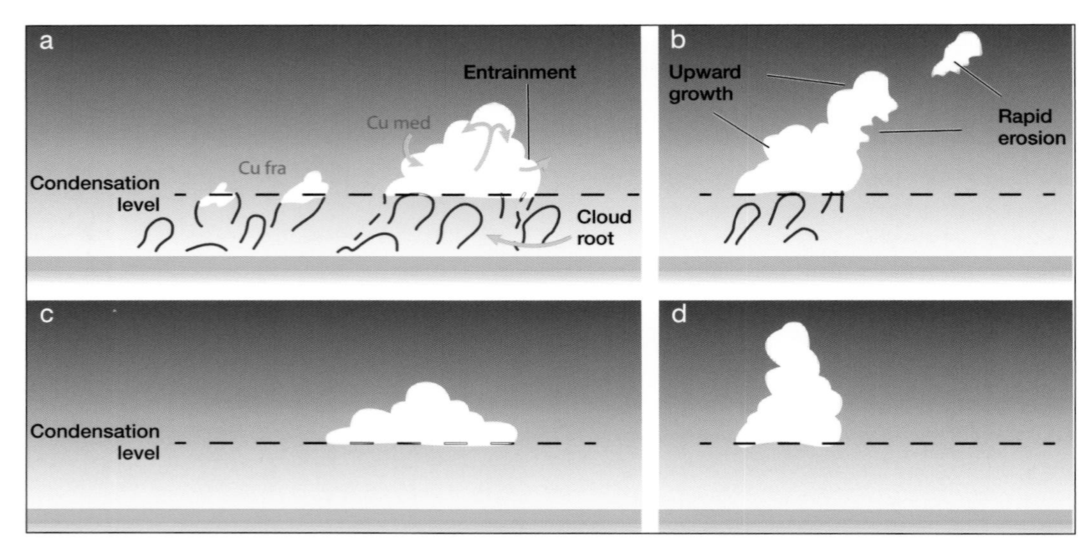

RIGHT Cumulus clouds appear when individual bubbles (or groups of bubbles) rise above the condensation level (a). Wind shear may cause them to dissipate rapidly (b). With low instability, the resulting clouds will be shallow (c), but with greater instability, considerable upward growth will occur (d). *(Ian Moores)*

The deepest clouds of the Cumulus genus, Cumulus congestus, may give rise to precipitation in the form of rain. This precipitation has been formed by coalescence of cloud droplets ('warm rain') and occurs only in very vigorous, deep clouds. At temperate latitudes, sufficiently deep clouds may form in late spring and summer, but in the tropics they occur all the year round. The showers that they produce consist of large raindrops, but are generally of short duration. When freezing (glaciation) occurs in the cloud tops, they become Cumulonimbus, and these may give much heavier and longer outbreaks of rain ('cold rain') or hail.

When Cumulus clouds reach a stable layer, unless they are particularly vigorous they may be unable to break through it, so they tend to accumulate beneath the layer and spread out sideways, giving a layer of Stratocumulus cumulogenitus (Sc cugen) or, at a higher level, Altocumulus cumulogenitus (Ac cugen). The resulting layer is often thicker (and thus darker) than 'ordinary' Stratocumulus (or Altocumulus) stratiformis, and may even be so thick as to reduce heating of the ground and inhibit the formation of further Cumulus clouds.

When a stable layer is encountered just above the condensation level, shallow Cumulus humilis are formed. These are often particularly noticeable ahead of an advancing warm front, when wisps of Cirrus are seen high overhead.

When there is a single source of heating, a whole series of clouds may form downwind of the source. More frequently, however, and particularly when the upper air is subsiding (such as under anticyclonic conditions), there may be a reasonable depth of unstable air beneath the inversion. The Cumulus clouds that form are somewhat deeper than Cumulus humilis and they occur as long rows extending downwind. Air is rising within the clouds and descending in the gaps between the lines of cloud. Such Cumulus radiatus (cloud streets) are usually

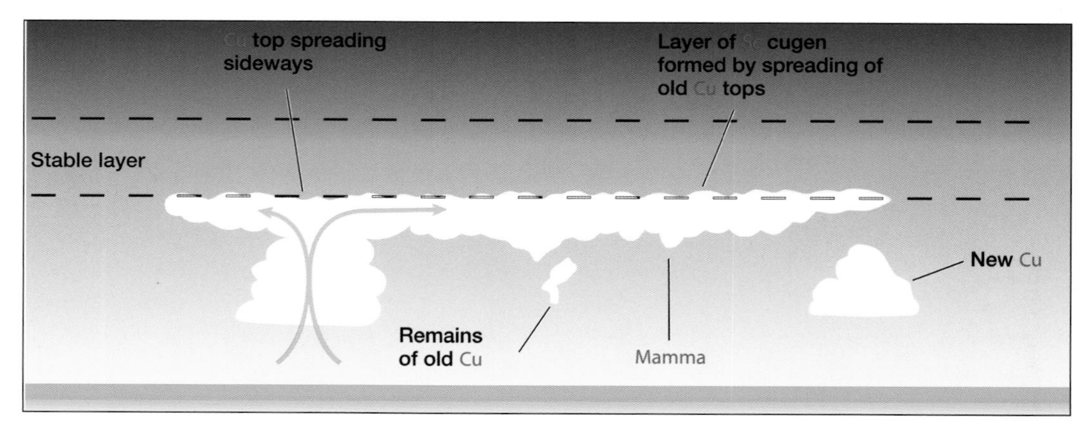

RIGHT The formation of Stratocumulus (or Altocumulus) cumulogenitus when rising Cumulus clouds encounter a stable layer. The layer tends to erode by mixing at its lower surface, sometimes producing distinct bulges or mamma. *(Ian Moores)*

LEFT A layer of Stratocumulus forming from Cumulus, with a separate layer of low Altocumulus above. *(Author)*

BELOW LEFT An extensive field of Cumulus humilis beneath Cirrus thickening into Cirrostratus ahead of a warm front. *(Author)*

ABOVE Cumulus cloud streets (Cumulus Radiatus), streaming eastwards on a westerly wind, with jet-stream Cirrus above. *(Author)*

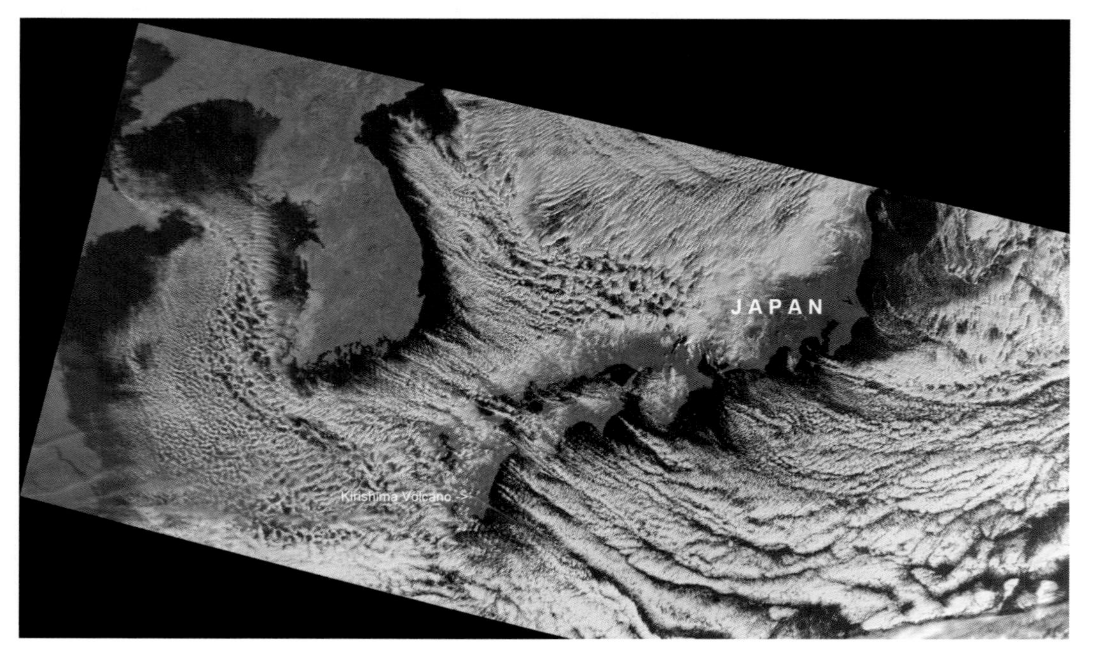

LEFT Extensive cloud streets forming in the generally north-westerly wind over Korea and Japan. A narrow trail of fumes from the Kirishima volcano is also visible. *(Eumetsat)*

ABOVE Some shading on the cloud elements and clear blue sky visible between the elements show that this cloud is high Altocumulus perlucidus. *(Author)*

BELOW Cirrocumulus floccus with virga, and some encroaching Cirrostratus. *(Author)*

separated by approximately two or three times the depth of the unstable layer. Vast areas covered by cloud streets may often be seen in satellite images of the North Atlantic, where cold air is flowing off the Greenland ice sheet and out over the relatively warm oceanic waters.

Convection is involved in the break-up of Altostratus or Stratus into Altocumulus and Stratocumulus perlucidus (in particular). When the top of a cloud layer radiates heat away to space, the lapse rate within the cloud becomes greater than the SALR, such that descending currents of cold air are created that produce gaps between the individual cloud elements. Similarly, a form of convection is one proposed cause for the formation of mamma: the pendulous masses of cloud that are particularly prominent beneath overhanging Cumulonimbus anvils, but which also occur with other cloud genera.

Convection is also involved in higher-level clouds, particularly in Altocumulus. Both Altocumulus castellanus (Ac cas) and Altocumulus floccus (Ac flo) are indications of instability in the middle layer. They are commonly accompanied by virga, where precipitation is falling from the clouds but evaporating before it reaches the ground. These forms are an indication that severe weather (Cumulonimbus and thunderstorms) are likely to develop in the near future. If Cumulus clouds (particularly Cumulus congestus) reach the same level, the instability present may result in explosive upward growth, leading to severe weather.

Because of the lower temperatures existing at the highest level, which mean that most clouds consist of ice crystals, rather than water droplets, Cirrus castellanus and Cirrus floccus

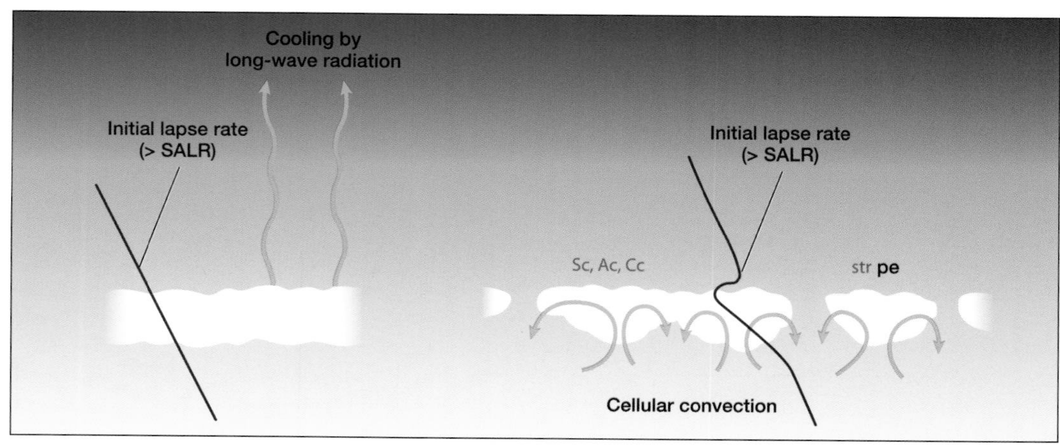

RIGHT A layer of stratiform cloud may lose heat from its upper surface by radiation to space. This cooling creates descending currents, which break up the layer into individual elements, producing the perlucidus variety. *(Ian Moores)*

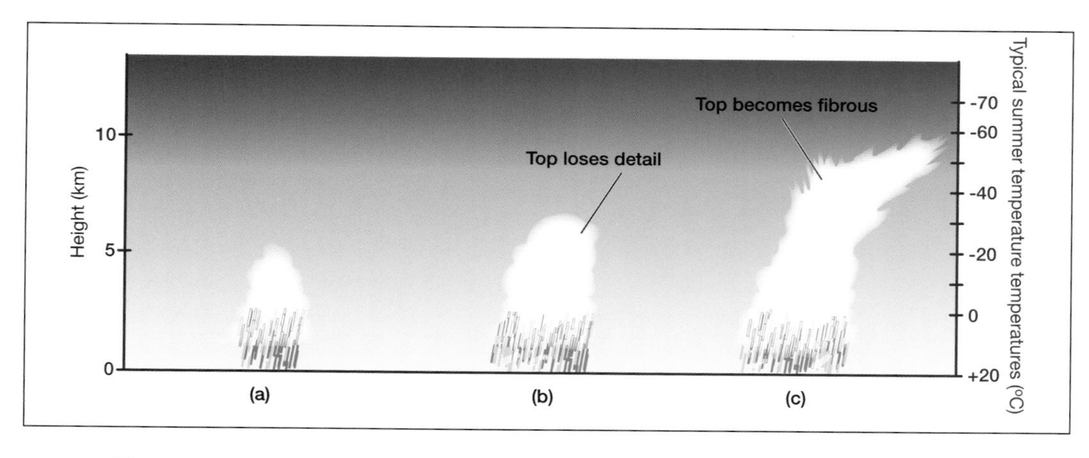

LEFT A Cumulus congestus cell (a) evolves into Cumulonimbus calvus (b) when glaciation begins in its top. It becomes Cumulonimbus capillatus (c) when extensive Cirrus wisps become visible. *(Ian Moores)*

are rare. They do, however, occur occasionally, and generally resemble the Altocumulus forms.

Cumulonimbus clouds

Cumulonimbus clouds arise from large Cumulus congestus. It is often difficult to recognise when the transition takes place, but the tops of the congestus clouds lose their hard, 'cauliflower-like' definition and become diffuse – a sign that glaciation has begun in the coldest parts of the cloud. (This is usually when the temperature is below -20°C, although it may sometimes occur at higher temperatures, but always below -10°C.) The rising cell has become Cumulonimbus calvus. The next stage (Cumulonimbus capillatus) is somewhat easier to detect, because the top of the cloud becomes fibrous in appearance, with extensive Cirrus. This may form a dense mass of fallstreaks or Cirrus spissatus (Ci spi cbgen).

When a rising tower reaches a stable layer (often the tropopause), the cloud spreads out sideways into an anvil, often with mamma beneath the overhanging cloud. If there is considerable vertical wind shear, this tends to encourage the cloud's growth, and the top of the cloud may extend as a dense mass akin to Altocumulus or Nimbostratus. Unlike those water-droplet clouds – which rarely give precipitation – the base will appear diffuse, consisting as it does of falling ice crystals.

An individual cloud cell has a limited lifetime, but frequently cells arise in close proximity to

BELOW A detached anvil. The parent Cumulonimbus has largely dissipated, leaving a large Cirrus plume and some lower cloud behind. *(Author)*

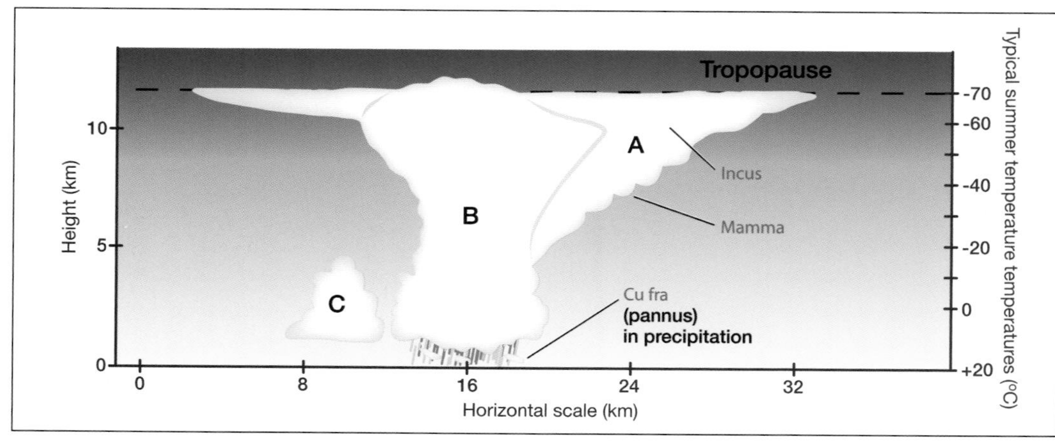

LEFT A simple Cumulonimbus cluster that has reached the anvil stage (Cb cap inc). A: Remains of old cell; B: Second, growing cell; C: First stage of new cell. *(Ian Moores)*

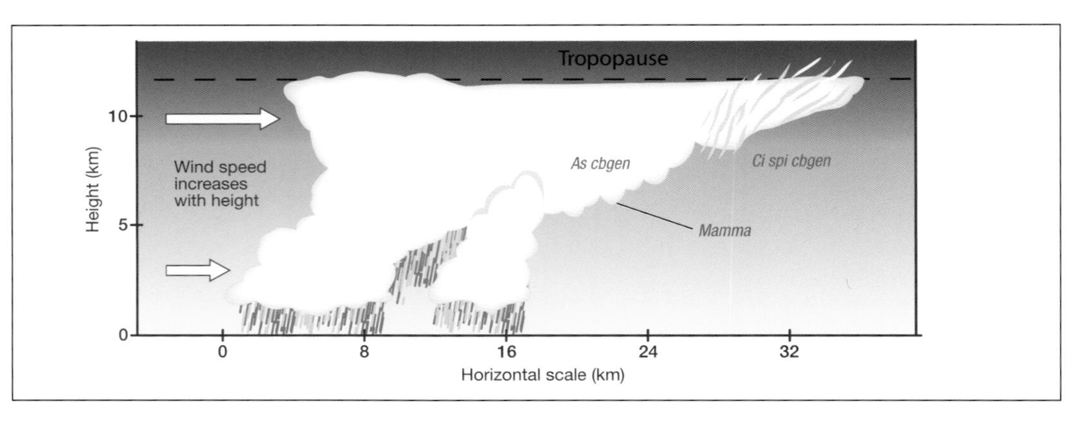

RIGHT **When there is considerable vertical wind shear, numerous cells may give rise to a mass of high, thick cloud that resembles Altostratus or Nimbostratus.** *(Ian Moores)*

one another, and cluster into a single system. This is particularly the case with vigorous Cumulonimbus cells, which draw air into the cloud at low levels, initiating the growth of new, closely adjacent cells. These follow the same pattern of growth, and if the process continues for some hours the succession of cells eventually contributes to the massive, single system that may cover a wide area of ground. Such multicell storms and the related supercell storms are described later.

When convection ceases, Cumulonimbus clouds gradually decay, but may leave various fragmented clouds behind them at various levels. The most obvious feature is often the remnant of any anvil, which persists as a mass of Cirrus spissatus (Ci spi cbgen). This 'detached anvil' may remain for many hours after the Cumulonimbus cell or cells have completely disappeared. Similarly, a Cumulonimbus may decay, leaving patches of Altocumulus (Ac cugen) and smaller patches of Cumulus (usually Cu med).

Stratiform clouds

Stratiform clouds form in stable air, but are associated with gentle uplift (unlike convective clouds) and weak vertical motion. They often cover large areas but are limited in their depth. The formation of mist and fog is described elsewhere, but some Stratus develops from fog that has risen away from the ground with the onset of daytime heating. Such Stratus may sometimes persist throughout the day.

As described earlier, Stratus may form through orographic uplift and often shrouds the tops of hills and mountains, when nearby lowland areas are clear of cloud. Another form of Stratus, the accessory cloud pannus, is created when precipitation from a higher cloud (Nimbostratus, Altostratus, Cumulus or Cumulonimbus) evaporates in the air beneath the cloud, which is therefore both moistened and cooled. Turbulence results in fragmented wisps of Stratus cloud being formed between the main cloud mass and the ground.

Stratus also forms like advection fog, where humid air blows across a cold surface, whose temperature is below the dew point. Whether fog or Stratus is formed depends on the strength of the wind and on the temperature difference between air and ground. As with radiation fog, advection fog forms at low wind speeds – below about 5 knots (9kph) – and Stratus at higher speeds – 5–10 knots

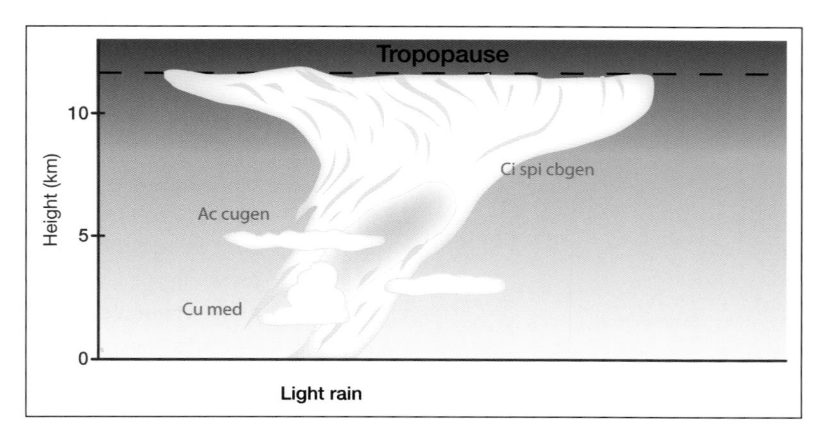

BELOW **When a major Cumulonimbus ceases to be active, it may leave behind various patches of cloud, most notably a large 'detached anvil', consisting of Cirrus spissatus.** *(Ian Moores)*

BELOW RIGHT **Overnight radiation fog formed over a lake starting to lift into Stratus cloud.**

(approximately 9–18kph). At higher wind speeds there is a tendency for mixing to be spread throughout a deep layer and for the decrease in temperature to be insufficient for cloud to form. Persistent low Stratus caused by low sea temperatures in the North Sea and moist easterlies is known as 'haar' in Scotland and north-eastern England, and this term is sometimes extended to advection fog, rather than true Stratus cloud.

Pyrocumulus

The term 'pyrocumulus' is a rather ugly word – a combination of both Greek and Latin elements – for a cumuliform cloud that has been created by the heat from deliberate or accidental fires. Now that stubble burning has been banned in many countries, the most common source of these clouds is a wildfire. Pyrocumulus from major wildfires are even able to grow into Cumulus congestus or Cumulonimbus. In some cases, when there has been little wind, the precipitation from the cloud has even been known to partially quench the fire that produced the cloud in the first place. Instances are also known of fires giving rise to major thunderstorms, lightning strikes from which have actually initiated further fires.

Contrails

Contrails (condensation trails) left by aircraft are extremely familiar. On closer examination, they offer information about current and forthcoming conditions. The exhaust from the aircraft's engines is sucked into the centre of the otherwise invisible vortices that are shed by each wing tip. All contrails therefore actually consist of twin trails of condensation, and these trails often develop downward bulges, somewhat resembling the mamma that occur under certain clouds. These bulges arise from the interaction between two forces: air is being forced downwards – that is what keeps the aircraft up – and, at the same time, the air is being heated by the exhaust. The result is a form of convection, very similar to the 'upside-down convection' that is thought to explain the bulges in mamma.

When the air at the flight altitude is dry, contrails dissipate quite quickly, but when it is humid they may persist for an extremely long time. Persistent contrails that spread out sideways are one of the signs of warm humid air encroaching aloft – *ie* that a warm front is approaching. Sometimes contrails may spread out so widely that they cover nearly all the sky, and reduce the amount of heat reaching the ground. The air at normal flight altitudes is cold, so persistent contrails tend to become glaciated, appearing like wisps of

LEFT Dense Cumulus congestus pyrocumulus clouds looming above the smoke from a major wildfire in northern Florida. *(Durham Garbutt)*

Cross references

Advection fog p.103
Dew point p.164
Mist and fog p.102–104
Pannus p.74
Radiation fog p.102

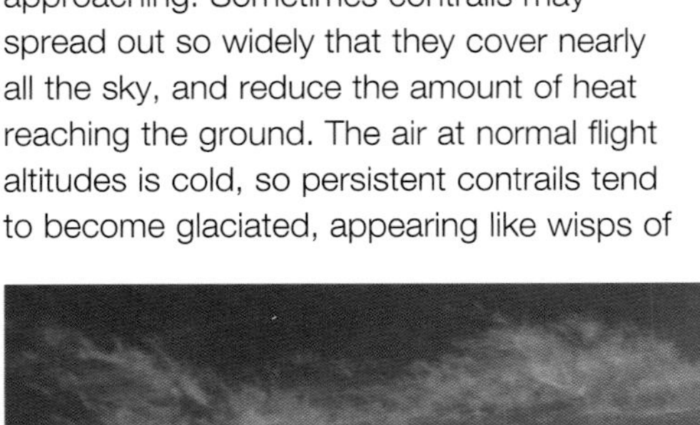

LEFT Side view of a contrail, showing the 'loops' beneath the main trail, caused by air being forced downwards by the action of the aircraft. *(Author)*

LEFT Persistent contrails slowly spreading across the sky in the humid, warm upper air, ahead of an approaching warm front. *(Author)*

LEFT Three glaciated contrails, which are slow to disperse in the cold upper air. *(Author)*

RIGHT A distrail ('dissipation trail') in high Altocumulus, where passage of the aircraft has induced glaciation, leading to the glaciated trail of virga to the right of the clear lane in the cloud. *(Author)*

Cirrus, and the falling ice crystals often produce virga-like trails.

Distrails

When an aircraft flies at the same altitude as a thin layer of cloud, it may produce a 'distrail', an elongated gap in the cloud cover. There appear to be several mechanisms by which this can occur: the aircraft may mix warmer air down into the cloud layer, dispersing the cloud droplets; or the heat from the engines may have a similar effect. In many cases, however, especially when the cloud droplets are supercooled, the introduction of tiny particles in the exhaust may cause the droplets to freeze. The resulting ice crystals fall out, and may sometimes be seen as a line of cirriform 'cloud' beneath the gap in the cloud layer.

RIGHT Passage of an aircraft through this thin layer of Altocumulus has probably initiated glaciation in the cloud. The ice crystals have fallen out, creating the fallstreak hole and producing a dense patch of Cirrus beneath it. *(Author)*

LEFT Jet-stream Cirrus forms highly distinctive bands of cloud across the sky. *(Author)*

Cross references

Billows p.84
Jet streams p.25
Low-pressure systems p.39
Wind shear p.166

Fallstreak holes

Clouds containing fallstreak holes – also known as 'hole-punch clouds' – are usually fairly thin layers of Altocumulus or Cirrocumulus, in which more or less circular holes appear. Such holes were unknown until aircraft began flying at the appropriate altitudes (in the 1920s and 1930s), and appear to result from a similar mechanism to distrails. The passage of an aircraft through the cloud layer seems to be sufficient to initiate glaciation in the cloud, which spreads out over an approximately circular area. Again, as with distrails, Cirrus-like wisps of cloud (fallstreaks) are often clearly visible within the hole. In fact, these normally lie just below the main cloud layer.

Jet-stream Cirrus

As already mentioned, the extreme wind shear that exists on the boundaries of jet streams tends to create distinct billows, lying across the flow. Because of the high speeds within a jet stream, however, Cirrus clouds that form within it are frequently drawn out into long streaks downwind. Such Cirrus radiatus is extremely characteristic of jet-stream Cirrus. Although superficially similar to contrails, the density and structure of jet-stream Cirrus mean that it is rarely confused with other cloud formations.

Because of their altitude, it is often difficult to detect any obvious motion in these clouds, but their appearance is so distinctive that they usually indicate the direction of airflow within the jet. Bearing in mind the relationship of jet streams to surface frontal systems, they provide a clue as to the way in which the weather is likely to develop. A north-westerly jet (*ie* one where the motion is approximately towards the south-east), often lies ahead of, and roughly parallel to, the warm front of a depression. When the lower wind direction – perhaps shown by Cumulus humilis clouds – is south-westerly, the 'crossed winds' are a clear indication of an approaching warm front. In a similar fashion, a south-westerly jet tends to lie on the polar side of a cold front.

Chapter 7

Precipitation and deposits on the ground

The term precipitation is used for all forms of liquid or solid water that originates in the atmosphere and falls to the surface. It therefore includes: drizzle, rain, freezing rain, ice pellets, ice crystals, hail and snow. It does not include those forms that are either suspended in the atmosphere (clouds, mist, fog) or deposited directly onto the surface: dew, frost or rime.

The different forms of particle (water droplets, ice crystals etc) occur at different heights and in different forms of cloud. In the highest étage (in the Cirrus family of clouds), for example, there are ice crystals, snowflakes, snow pellets and supercooled water droplets. The clouds normally associated with specific precipitation are given in the following table.

Name	*Clouds that may give the precipitation*
Rain	Ns, As, Sc str op, Ac flo, Ac cas, Cu con, Cb.
Drizzle	St, Sc str op.
Freezing rain or drizzle	As for rain and drizzle.
Snowflakes	Ns, As, Sc str op, Cb.
Sleet	As for snowflakes (in Britain).
Snow pellets (soft hail; graupel)	Cb in cold weather.
Snow grains	Sc str op, St in cold weather.
Ice pellets (sleet in North America)	Ns, As, Cb.
Small hail	Cb.
Hail	Cb.
Diamond dust	St, Ns, Sc str op, or even from clear air.

Forms of precipitation

Name	*Description*
Rain	Drops generally with diameters > 0.5mm.
Drizzle	Drops with diameters < 0.5mm.
Freezing rain or drizzle	Rain or drizzle that freezes on contact with a cold surface.
Snowflakes	Loose collections of stellar or dendritic ice crystals.
Sleet	In Britain, partially melting snowflakes or combined snow and rain; in North America, ice pellets.
Snow pellets (soft hail; graupel)	Opaque, white grains of ice, generally spherical, with diameters 2–5mm.
Snow grains	Tiny, white, opaque grains of ice; diameters generally < 1mm.
Ice pellets (sleet in North America)	Translucent or transparent ice pellets, spherical or irregular in shape, with diameters up to about 5mm.
Small hail	Translucent ice pellets, consisting of snow pellets covered in a layer of ice; diameters less than 5mm.
Hail	More or less spherical balls of ice, often consisting of clear and opaque layers; diameters 5–50mm or even larger, sometimes combined into larger pieces.
Diamond dust	Minute crystals of ice that appear to be suspended in the air.

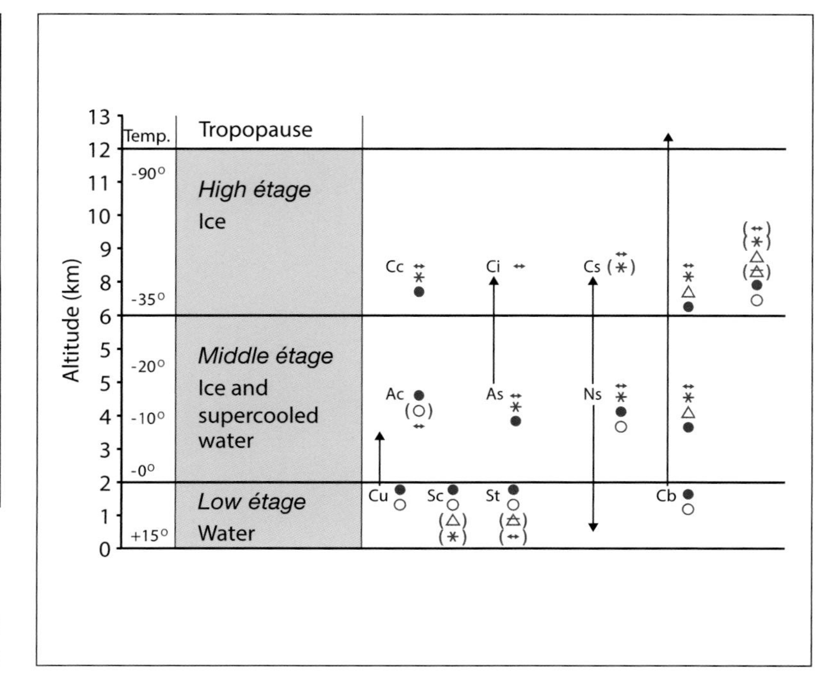

RIGHT Cloud types, their range in altitude and typical forms of precipitation. *(Ian Moores)*

ABOVE **A precipitating Cumulonimbus cloud off Florida. The softness at the top of the cloud indicates that glaciation has commenced, and that the cloud is producing 'cold rain'.** *(Durham Garbutt)*

Liquid precipitation

Rain

Technically, raindrops are generally defined as having diameters greater than 0.5mm (although some smaller drops are called rain if they are widely scattered). Drizzle consists of tiny drops with diameters less than 0.5mm. These sizes may be compared with the diameters of actual cloud droplets, which range from about 0.002 to 0.1mm. A typical raindrop has a diameter of 2mm, but its volume is one million times greater than that of a typical cloud droplet (with a diameter of 0.02mm).

RIGHT **Tall, summer-time Cumulus congestus over the English Channel. The ragged bases indicate that precipitation, created by the coalescence process and thus 'warm rain', is imminent.** *(Author)*

Cloud droplets arise through condensation on condensation nuclei, but the latter are so abundant throughout the atmosphere that although the condensation process could produce larger droplets, given sufficient time, the presence of so many nuclei means that vast numbers of small droplets are produced rather than fewer, larger ones. Condensation alone appears to be able to produce drizzle droplets (about 0.5mm in diameter). Anything larger would swiftly fall out of the saturated region, bringing growth to a halt.

Two processes are able to create the larger, true raindrops. In a mixed cloud below 0°C, containing both supercooled water droplets and ice particles, water vapour is deposited on the ice particles at the expense of the droplets. The process removes water vapour from the air, so the water droplets evaporate, whereas the ice particles grow into larger crystals. These fall out and eventually melt into raindrops. The whole process is known as the Bergeron-Findeisen process (or, occasionally, as the Wegener-Bergeron-Findeisen process).

This gives rise to what meteorologists sometimes refer to as 'cold rain', that is, rain that originates through a freezing process as ice crystals, which then subsequently melt. This process accounts for rain from clouds in which glaciation occurs, *ie* Cumulonimbus, but cannot explain rain from clouds in which the temperature is always above 0°C, in the so-called 'warm clouds'. Such clouds are found year-round in the tropics, and in summer at middle latitudes.

We need a second process to account for 'warm rain'. This is now generally accepted to be coalescence, that is, through collisions between droplets that then merge. This process occurs within clouds that have vigorous convection, *ie* Cumulus congestus and Cumulonimbus. Water vapour condenses on suitable nuclei, and some droplets become larger than the others. The updraughts within the cloud are unable to support the larger, heavier droplets, which begin to fall towards the ground. As they do so they collide with smaller droplets, merging with them and becoming still larger. If droplets reach a diameter of about 6mm, they break up into smaller droplets, setting a limit to the size that raindrops may attain.

Layer clouds and precipitation

Stratus and Stratocumulus are shallow clouds (normally their depth is less than 1,000m) and possess weak (if any) convection. Their water content is also very low, and their temperature is rarely below -10°C, so they do not generally contain any ice crystals. Any precipitation is by coalescence, so it is very light and in the form of drizzle droplets or very fine rain. If such clouds are forced to rise over high ground the additional cooling may produce somewhat heavier fine rain. To anyone within the cloud it is a form of 'Scotch Mist'.

Occasionally, the temperature of the tops of stratiform cloud may drop below -10°, and a few small ice particles may form. These grow by the Bergeron process and may result in precipitation in the form of tiny snow grains (diameter < 1mm), which may reach the surface if temperatures are below freezing down to the surface.

Under very cold conditions, the minute crystals known as diamond dust may also be produced in Stratus, Nimbostratus, or thick Stratocumulus.

The slow uplift at frontal systems produces widespread, thick stratiform clouds (Nimbostratus and Altostratus) that are predominantly responsible for prolonged and usually heavy precipitation that accompanies depressions. The temperature at the top of these clouds is often low enough for ice crystals to form, which then grow (in the Bergeron process) by accumulating supercooled water droplets. The resulting snowflakes may then melt into rain. In winter, when surface temperatures are below freezing, the particles remain frozen and major snowfall may then occur, at both warm and occluded fronts (in particular).

Solid precipitation

Snow

As we have seen, rain often begins as ice crystals formed high in clouds that then melt at lower altitudes into raindrops. The crystals themselves have various forms, including the intricate, six-sided crystals commonly regarded as 'snowflakes'. In fact, true snowflakes consist of many individual crystals that have collided with one another, generally in clouds where the temperature is just below 0°C. Under such circumstances the thin film of water on their surfaces then freezes, locking them together. At lower temperatures the individual crystals tend to remain separate.

The type of snow that reaches the ground entirely depends on temperature. At low temperatures, the snow is in the form of small, separate crystals. This forms a layer of the 'dry' or powder snow favoured by skiers. However, when the temperature is near freezing large flakes of snow will fall. This 'wet snow' poses a particular problem for transport authorities because, unlike dry snow, it cannot be blown out of the way. It melts under pressure, but

ABOVE Heavy snowclouds and snowfall over Zermatt in Switzerland. *(Dave Gavine)*

BELOW At high altitudes, the low temperatures result in deep powder snow, consisting of innumerable tiny, individual ice crystals. *(Author)*

RIGHT When first deposited on the ground or other objects, snow traps large amounts of air between the individual snowflakes. *(Author)*

RIGHT Mixed clouds (including Altocumulus lenticularis) over the Aletsch glacier in Switzerland. The glacier's ice consists primarily of compacted snow. *(Author)*

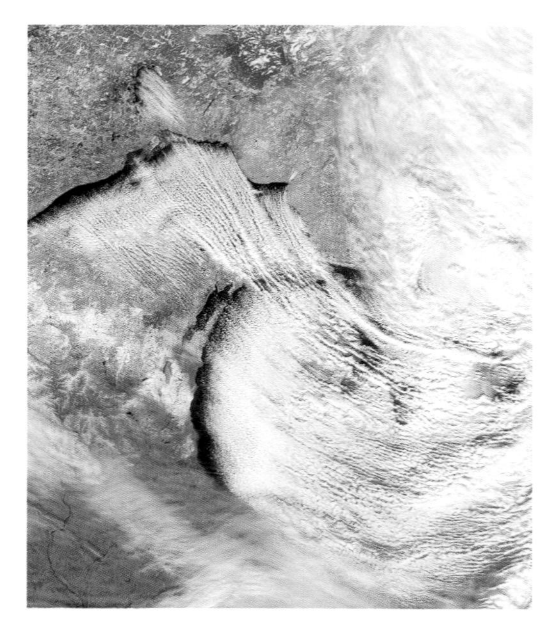

RIGHT On 5 December 2000, a cold north-westerly airstream across Lakes Superior and Michigan gave rise to numerous cloud streets – particularly marked over Lake Superior – as it flowed across the relatively warm water. Heavy 'lake-effect snow' was deposited over almost the whole of Michigan State. *(NASA)*

RIGHT Utility poles that were snapped by the weight of ice deposited during the ice storm of January 1998 over Canada and New England. *(Durham Garbutt)*

reforms as ice as soon as pressure is removed, causing a major hazard on roads, railway lines and airport runways.

When first deposited, snow contains large amounts of air, trapped between the individual snowflakes. If undisturbed, over time, it gradually becomes compacted – particularly if subject to repeated cycles of slight thawing and re-freezing – and turns into ice. This is how glaciers develop in mountainous country.

When very cold air flows across relatively warm water, it may pick up large quantities of water vapour, which may then be deposited as heavy falls of snow on the leeward shore. The effect is often known as 'lake-effect snow' and is particularly common on the shores of the Great Lakes in North America. Buffalo, in New York State, on the eastern side of Lake Erie, is among the locations noted for extreme snowfall of this type.

The effect is not, of course, confined to lakes, and similar conditions and snowfall may occur that originate in moisture evaporating from seas or oceans. The resulting snow is sometimes known as 'ocean-effect' or 'bay-effect' snow.

Glaze

Sometimes, supercooled rain or drizzle droplets may reach the ground. If the air and objects at the surface are below 0°C, the droplets freeze immediately on contact with any surface, leading to a coating of glaze, giving rise to the treacherous layer known as 'black ice' on road surfaces. Occasionally, such occurrences of glaze are extremely widespread and severe, such as the 'ice storm' that affected large areas of Canada and New England in January 1998, which brought widespread disruption to electricity supplies as well as transport.

The lesser forms of frozen water are:

- **Sleet** – In Britain, partially melted snowflakes or a combination of snow and rain. In North America sleet is generally taken to be ice pellets, but the term is also (confusingly) sometimes applied to wet snow as in Britain.
- **Snow grains** – Tiny grains of ice about 1mm across, which are, in effect, frozen drizzle, and fall from shallow stratiform clouds.

- **Snow pellets** – Small, opaque grains of ice, formed when tiny supercooled droplets freeze together, trapping some air into the spaces between them. They are sometimes known as soft hail or graupel, and may be regarded as the initial forms of hail.
- **Ice pellets** – These originate as raindrops or melted snowflakes, which either arose in a warm layer of air or passed through one before falling into a layer with sub-zero temperature, where they froze.
- **Small hail** – Translucent pellets consisting of a core of snow surrounded by a layer of clear ice. They form when the cores have fallen into a layer of warmer air, accreting liquid droplets that then form a clear layer of ice.

Hail

Hailstones normally consist of alternating layers of clear and opaque ice. These are laid down in layers of air at different temperatures. The opaque layers occur when supercooled droplets freeze on to the initial particle, trapping air in the cavities between them. The clear layers are deposited when the particle falls into a layer with a temperature above freezing, when the liquid water covers the particle with a layer of water that subsequently freezes. The necessary conditions occur in deep Cumulonimbus clouds, where the main updraught is tilted rather than vertical. Frozen particles, created at the top of the cloud, grow during their descent, but then fall into the powerful updraft and are carried upwards again. This process may continue several times, until the hailstones are too heavy to be swept upward again and fall out of the cloud.

Hailstones may be produced by moderately active Cumulonimbus clouds, but the very largest are created in supercell storms, because supercells have extremely powerful updraughts that are able to lift very heavy hailstones. The largest individual hailstones are approximately the size of a grapefruit, and some Indian stones have weighed 1kg. Far larger hailstone aggregates occur when individual stones become frozen together. The largest known aggregate fell in China and weighed 4kg.

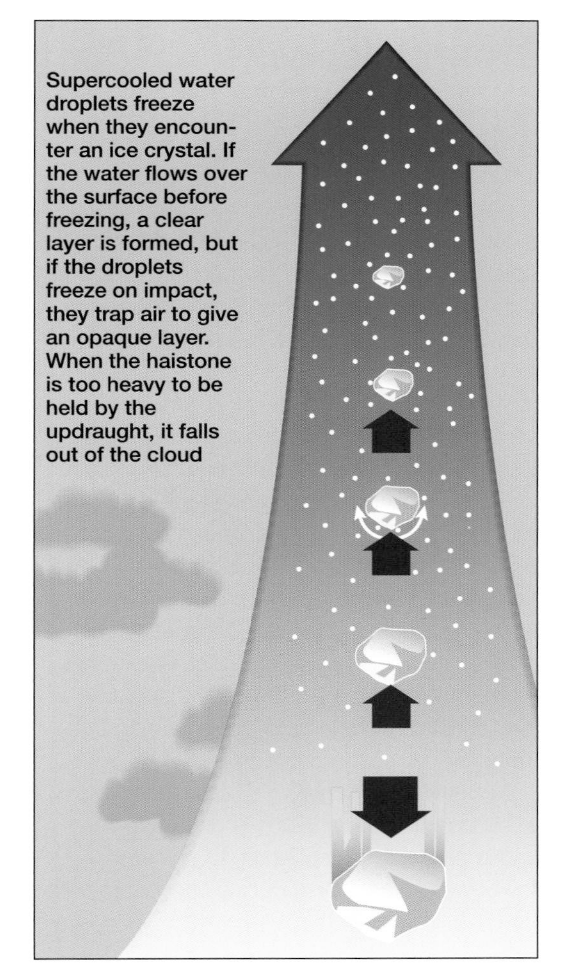

LEFT Supercooled water droplets freeze when they encounter an ice crystal. If the water flows over the surface before freezing, a clear layer is formed, but if the droplets freeze on impact, they trap air to give an opaque layer. When the hailstone is too heavy to be held up by the updraught, it falls out of the cloud. *(Ian Moores)*

Cross references

Cumulonimbus p.73
Diamond dust p.104
Ice crystals p.135
Supercells p.116

LEFT Some of these hailstones (which actually fell in California) have broken on impact with the ground, revealing their inner, layered structure. *(Steve Edberg)*

LEFT Heavy precipitation from a Cumulonimbus cloud falling over Brannenburg in the Inn Valley in Bavaria. The rain appears grey and the hailshafts are white. *(Claudia Hinz)*

Cross references
Dewbow p.133
Heiligenschein p.134

Ground deposits

The most common form in which water is deposited directly on to the ground or surface objects is dew. This forms whenever blades of grass, leaves or other objects are directly exposed to the sky and lose heat by radiation. As soon as the temperature drops to the dew point, dew droplets are deposited from the air. Generally the water vapour is derived from the soil, which, largely protected from direct exposure to the sky, remains warmer than higher surfaces. Dewdrops are generally small, being about 1mm in diameter. Larger droplets are often seen at the ends of blades of grass or leaves. These have not been deposited from the air, but have arisen through the transport of water from the roots of the plants into the leaves. Because the air is extremely humid the water cannot evaporate in the normal fashion from the leaves, so it forms large droplets (known as guttation drops) that are usually at least 2mm across.

Dewdrops, particularly those suspended on horizontal spiders' webs spun between blades of grass, may give rise to coloured dewbows. An additional optical effect seen with a large dew-covered expanse of grass is the heiligenschein, a bright halo seen around the shadow of the observer's head.

Hoar frost

If the air temperature drops below 0°C before dew forms, the water vapour is deposited directly on to the surface of objects as ice crystals, without passing through a liquid phase. The resulting deposit of soft ice crystals is known as hoar frost. In some cases the deposit may be so thick that, from a distance, it appears like snow.

Sometimes dew will form on surfaces, and the temperature later drops below freezing. Generally the small droplets will not freeze immediately but will become supercooled. This is a common stage in the development of the fern-like patterns often seen on windows. Once a few initial ice crystals have formed, they tend to grow at the expense of the water droplets as water vapour is lost by the droplets and deposited on the ice crystals, which gradually grow intricate patterns across the glass. It is sometimes possible to see a clear gap between the dew droplets and the growing ice-crystal layer.

RIGHT A deposit of hoar frost that is particularly heavy where the gound has been exposed to the sky and thus lost heat though direct radiation to space. *(Author)*

FAR RIGHT TOP Frost has been deposited on this spider's web, but has not been heavy enough to break any of its strands. *(Dave Gavine*

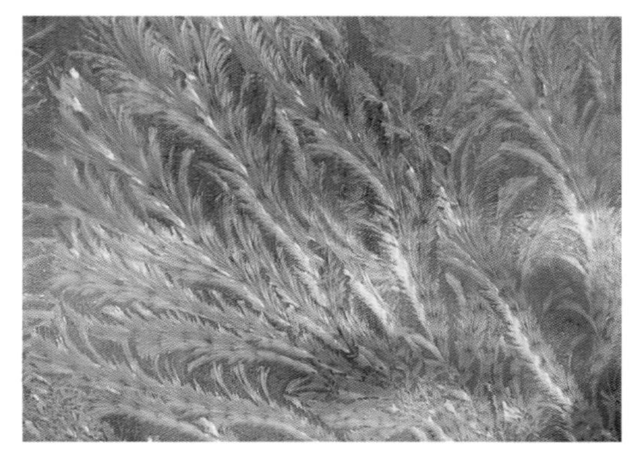

FAR RIGHT Dendritic frost trails formed by supercooled water droplets freezing onto random initial ice crystals, and growing into intricate patterns. *(Author)*

Rime

Although superficially resembling hoar frost, rime icing occurs through a completely different process. It is deposited when supercooled fog droplets come into contact with solid objects that are below freezing. The droplets freeze almost immediately on contact with any object, leading to a white ice deposit.

Because of the way in which it forms, rime tends to occur on the windward side of any object, so that if there is a significant drift of the air laden with supercooled droplets long, icy 'feathers' may be produced, pointing into the wind. Under still conditions, however, long 'needles' of ice are often deposited around the edges of leaves or other objects.

As with glaze, rime deposits may occasionally be very heavy and cause some damage, particularly to the twigs and branches of trees and shrubs. High-altitude meteorological stations often suffer from extreme heavy rime deposits that prevent the instruments from recording accurate data. At manned stations the rime is sometimes removed by hand, but at automatic stations various forms of heating are used to try to keep the instruments free from the build-up of ice.

ABOVE Trees coated in rime, deposited by supercooled droplets in fog. *(Wikipedia)*

FAR LEFT The rime on these guy wires was created by the slow drift of air containing supercooled water droplets up the side of a mountain, so the 'feathers' point downwards at an angle to the horizontal. *(Author)*

LEFT Heavy rime on the instruments at the meteorological station on the Wendelstein mountain (1,838m) in the Bavarian Alps. *(Claudia Hinz)*

Chapter 8

Visibility

ABOVE A deep blue sky in the dry air high in the Swiss Alps at the Jungfraujoch Observatory. *(Author)*

BELOW The water vapour present in humid air scatters the longer wavelengths of light, causing distant objects to appear paler and thus giving rise to aerial perspective. *(Author)*

The transparency of the air depends on the type of particles suspended in it. Because the air is rarely completely dry, even over major deserts, a certain amount of water vapour is always present. At very high altitudes where the air is dry and free from suspended particles, the sky appears dark blue. At lower altitudes, any water vapour tends to scatter light so that the colour of the sky itself becomes a paler shade of blue, and distant objects become lighter in tint, giving rise to the aerial perspective that we unconsciously use to judge distance.

The transparency of the atmosphere is also affected by any particles suspended in it. Obviously visibility is greatly affected (and may be reduced to nothing) in a dust or sand storm, in torrential rain or in thick, blowing snow. There are, however, three common forms of lesser obscuration:

- Mist
- Fog
- Haze

The first two of these arise from water droplets suspended in the atmosphere. By international agreement, the term 'mist' is defined as allowing objects 1km or more away to be seen, and the term 'fog' is used when objects closer than 1km cannot be seen. However, for practical purposes, especially with regard to road traffic safety, weather forecasts for the general public use the term 'fog' when visibility is reduced to less than 200m.

The two most common forms of fog are:

- Radiation fog: occurring when the surface radiates heat away to space, cooling the air in contact with it.
- Advection fog: created when humid air flows over a cold surface and is cooled to the dew point.

Radiation fog

Radiation fog (or mist) is likely when several conditions apply:

- The sky should be clear, allowing radiation to escape into space.
- The air during the evening should be moist. (This is most likely to occur during autumn and winter seasons.)
- There should be adequate time for the air to cool to the dew point. (Again this is most likely to apply during autumn and winter.)
- There should be a light wind, generally less than 4 knots (7.5kph).

The last condition may seem somewhat surprising, but a light wind provides a degree of turbulence that mixes the air near the ground, causing the cooling to take place throughout a thin layer of air, whose depth generally lies between 15 and 100m. Very occasionally, thicker layers of fog (to a depth of 300m) may form. Humidity is likely to be high following rain during the day and in river valleys and near other bodies of water. Such locations favour the formation of radiation fog.

Radiation fog normally disperses if the wind rises, causing turbulence that mixes the humid layer with higher, drier air, or when sunlight begins to heat the underlying ground. Often the layer of fog will start to lift and become low-lying Stratus cloud, before disappearing completely. The warmth from the Sun may also initiate a valley wind, which may disperse the fog into small patches that are blown up the hillsides or mountainsides. Very early in the day a mountain wind may cause valley fog to flow downstream for a while before solar heating prevails and the fog starts to lift or a valley wind sets in.

Very shallow ground fog (1–2m deep) may sometimes form after late afternoon rain, when the air itself (rather than the ground) cools rapidly after sunset.

Advection fog

Advection fog generally forms when warm, moist air flows over a very cold surface (usually the sea or open ocean). The temperature of the latter must, of course, be below the dew point. Very extensive areas of sea fog may result and it is then carried by the wind (advected) on to neighbouring coasts. During the day, such fog may burn off over the land but persist over the sea. At night, when the land cools, the fog may return. Such sea fogs that invade the land are sometimes known as 'haar', a term originally used in Scotland and on the north-east coast of England, although this term really applies to low Stratus that comes in from the North Sea.

Similar advection fog may occur over land when warm moist air flows over a cold surface. This is particularly common when the land has been covered by snow and a thaw sets in, when the air temperature hovers around 0°C.

ABOVE Scotland's Great Glen, filled with valley fog – a radiation fog that had accumulated overnight. *(Dave Gavine)*

BELOW A satellite image showing extensive sea fog in the northern and southern North Sea, and in the English Channel. The fog has invaded parts of northern France, Belgium and East Anglia. *(Eurometsat)*

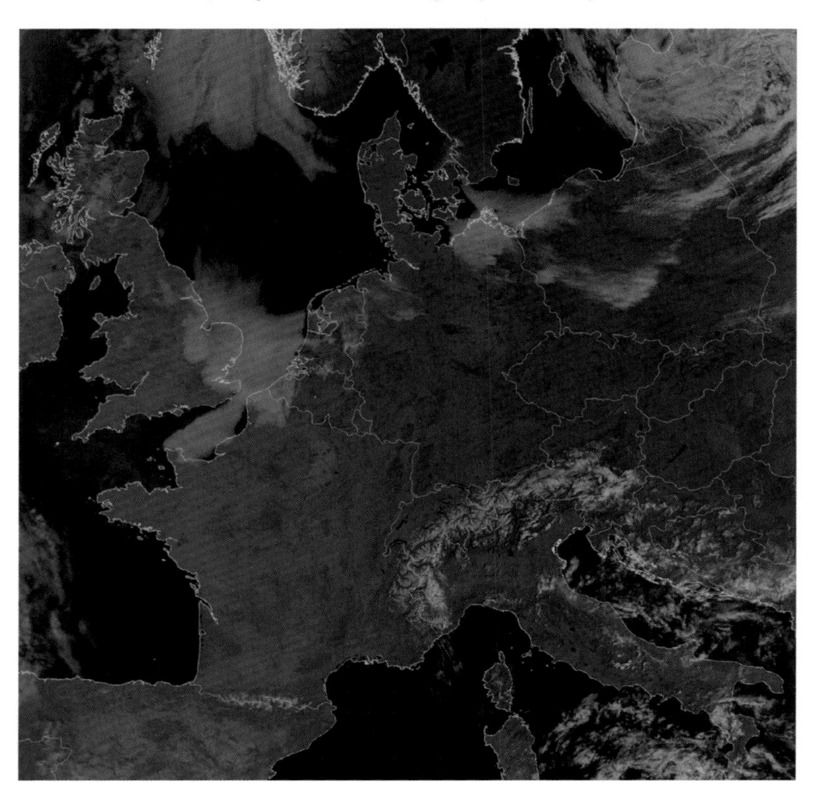

ABOVE A dense bank of sea fog slowly clearing with the warmth of the early-morning Sun. *(Author)*

Other forms of fog

In a reversal of the usual conditions, fog may occur when very cold air flows over warm water. Any water vapour rising from the water immediately condenses into tiny droplets, producing wreaths of 'steam' that may reach a height of as much as 50m. Such 'steam fog' is seen above ice-free rivers and lakes on very cold winter mornings. A more dramatic and extensive form, known as 'Arctic sea smoke', occurs in the polar regions when intensely cold air flows over open water.

Under extremely cold conditions, such as those found in Siberia, the Canadian Arctic or Antarctica, the water droplets in fog may freeze into tiny ice crystals, producing 'ice fog'. The crystals are so small that they rarely affect visibility, but instead glitter in the sunlight, when they are known as 'diamond dust'. Sunlight that is refracted and reflected by these tiny crystals produces brilliant halo effects and other optical phenomena, some of which have only ever been observed under these conditions.

Another form of fog is smog ('smoke fog'), where smoke or other pollutants provide plentiful, and extremely effective, condensation nuclei. The droplets that are created are smaller than normal fog droplets and they form more rapidly with declining temperatures. Because of the presence of so many condensation nuclei, these smogs also tend to persist longer that pure water fogs. This type of smog should not be confused with the obscuration that occurs when vehicle exhausts and other pollutants are altered by ultraviolet radiation, giving rise to the photochemical smog that typically occurs in major cities. On Hawaii, volcanic gases can produce a similar effect, giving rise to what is known locally as 'vog'.

Cross references

Condensation nuclei p.52–53
Dew point p.164
Mountain wind p.107
Precipitation p.95
Valley wind p.107

Haze

Haze is created by the presence of exceptionally small, dry particles that are suspended in the air, making distant objects slightly indistinct. The particles are normally so tiny that they scatter sunlight and, as a result, contribute the sunrise and sunset colours. Haze often builds up during the day and may form a distinct layer that usually appears brownish when seen against the light.

RIGHT The Sun setting behind a haze layer that built up during a hot summer's day. *(Author)*

Chapter 9

Local winds and effects

Because the wind plays such an important part in the weather around the world, there are dozens of local names for distinct winds that affect a particular area, which have specific characteristics and produce marked changes in the weather.

Sometimes the same term is applied to winds in different parts of the world. One example is 'doctor', which is applied to any wind that brings relief from hot, humid conditions. It is used to describe a welcome wind in north-western Africa, Southern Africa and Western Australia. Another is 'willywaw', which is used for a sudden gust that descends from the mountains to the sea. The term is used at both extremes of the Americas: in the Straits of Magellan in Patagonia and in the Aleutian Islands in the extreme north-west.

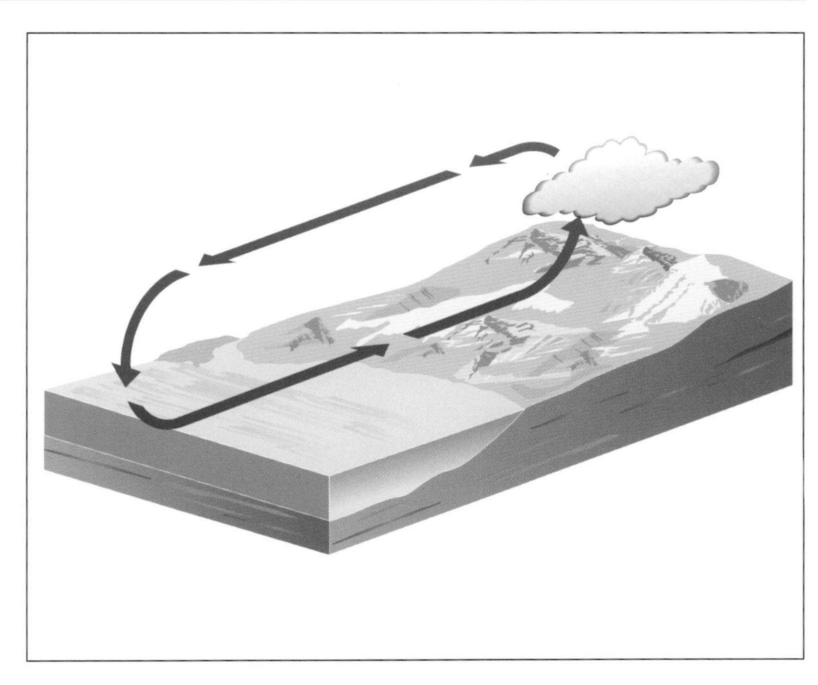

ABOVE During the day, air over the land heats more rapidly than the air over the sea, and the cold air flows inland as a sea breeze. *(Ian Moores)*

Localised winds

Quite apart from the large number of named winds that occur all over the world, there are other localised winds, produced by the particular conditions that prevail over an area of countryside, usually for part of a day. These localised winds have counterparts that are found in many different regions of the world and are basically caused by temperature differences between different areas of the land or sea. There are five main types:

- Sea breeze.
- Land breeze.
- Lake breeze.
- Valley wind.
- Mountain wind.

In addition to these specific types, the wind at any particular location is, of course, strongly influenced by local topography. A gap in a mountain barrier may experience extremely strong winds (known, for obvious reasons, as mountain-gap winds) when the air is funnelled through a small space. The levanter, for example, is concentrated between the Spanish highlands and the Atlas Mountains in North Africa. The mistral is similarly strengthened as it funnels down the Rhône valley. When the funnelling effect is particularly strong, the wind may be called a ravine wind. One of the most famous is the kosava, which occurs where the Danube breaks through the Carpathian Mountains to the east of Belgrade.

Sea and land breezes

A sea breeze is caused by the fact that during the day, land areas heat up faster than the neighbouring water. Thermals tend to build up over the land, rising and expanding, and the resulting pressure difference causes the cooler air over the sea to move inland as a sea breeze. On the coast and immediately inland, especially in spring and early summer when the sea and overlying air is cold, the sea breeze

ABOVE Clouds forming along the sea-breeze front where it has encountered a line of hills running parallel to the coast. *(Author)*

may bring sea mist inland. A bright, sunny morning may become a dull, cool afternoon.

As time progresses, the sea breeze penetrates further and further inland and may reach several tens of kilometres (or even more) from the coast. Generally there is a distinct sea-breeze front, where there is a definite change in temperature. Air is rising at this front and this may produce a line of cumuliform cloud (usually Cumulus) that gradually moves inland. If the sea-breeze front encounters higher land, such as hills or mountains, over which it is forced to rise, the clouds may become quite substantial Cumulus congestus, or even Cumulonimbus. These may produce rain over the hills, and even lead to significant thunderstorms.

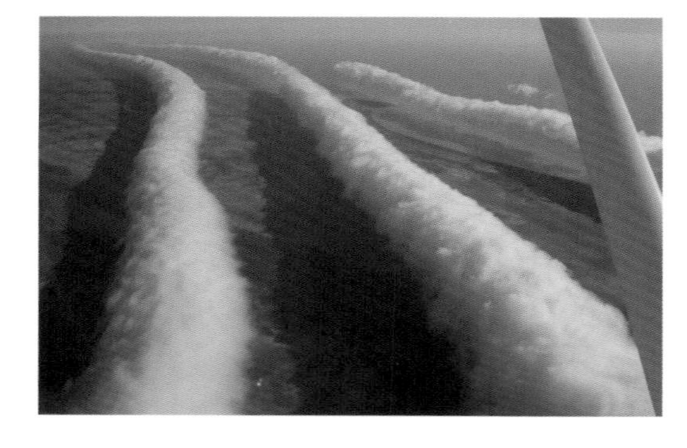

RIGHT A dramatic view of the Morning Glory near Burketown on an occasion when it displayed three distinct roll clouds. *(Wikimedia)*

RIGHT This photograph, taken on an early Gemini mission, clearly shows the lines of cloud along the land-breeze fronts on both sides of India. *(NASA)*

The sea breeze generally flows more or less at right angles to the coast, especially in the absence of any significant gradient wind. When the land is in the form of a peninsula, sea breezes may penetrate from the sea on both sides. Where they encounter one another along the spine of the peninsula, very significant growth of clouds (and major rainfall) may occur. This has been the cause of some of the major flooding events that have occurred in Cornwall, such as the dramatic Boscastle flooding in 2004, for example. (On a much larger scale, sea breezes from both sides of the Cape York Peninsula in Queensland, Australia, sometimes interact to produce the great roll cloud, which is the major feature of the squall line known as the Morning Glory, that propagates westwards across the Gulf of Carpentaria. The sudden jump in pressure when the squall arrives is sometimes followed by several distinct roll clouds.)

The air flowing inland as a sea breeze rises at the sea-breeze front and flows back out to sea at a moderate altitude. Sometimes a thin sheet of cloud develops in this airflow and may be seen gradually extending out to sea.

A land breeze is the night-time counterpart of the daytime sea breeze. The land cools more quickly when heating ceases at sunset (especially if the sky is cloud-free). The dense, cool air flows out to sea at a low level and such a land breeze usually commences around midnight and continues until dawn. As with the sea breeze, there is a front at the boundary between the warm and cool air, and this may result in a line of cumuliform cloud that gradually moves out to sea during the night. Satellite images obtained in the early morning often show the sea-breeze front as a line of cloud, many kilometres out to sea, that runs more or less parallel to the coastline. (A famous photograph obtained by Gemini astronauts shows land-breeze fronts on both sides of India, with the one over the Bay of Bengal hundreds of kilometres from land.) Just as with a sea breeze, a land breeze may produce Cumulonimbus clouds and even thunderstorms out over the sea.

RIGHT When heating ceases in late afternoon, air over the land cools rapidly, and then flows out to sea as a land breeze. *(Ian Moores)*

CENTRE Lake breezes arise under the same circumstances as sea breezes, although if the breeze encounters steep valley sides, clouds may build up very rapidly over the high ground. *(Ian Moores)*

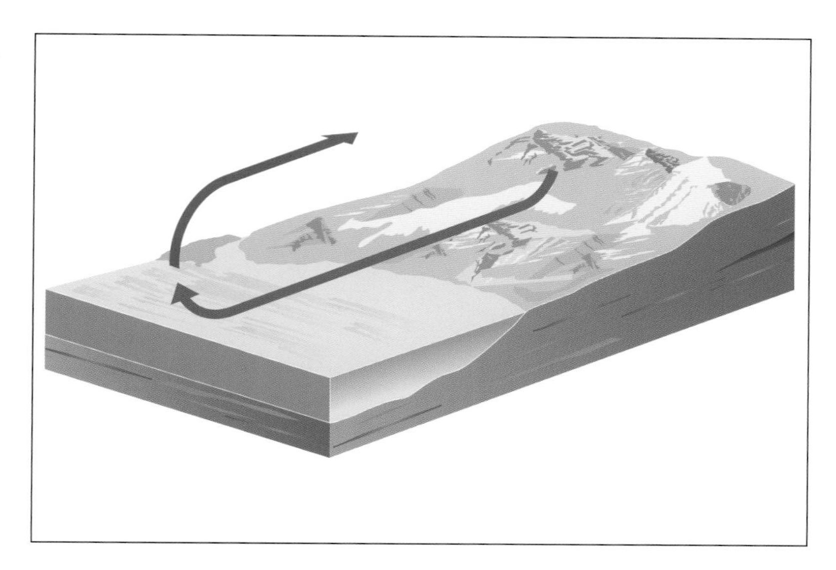

Lake breezes

Very large bodies of water, such as the Great Lakes in North America, may produce breezes that are as strong as sea breezes created next to the open sea. Smaller lakes also produce breezes by a similar mechanism, but here the strength of the breeze may be influenced by the depth of the water and by the surrounding countryside. Shallow lakes warm more quickly than deep ones, reducing the temperature contrast between the air over the lake and over the surrounding land, and giving a weaker breeze. The orientation of the lake also plays a part. If the lake lies in a distinct valley, one side of which is strongly heated by the Sun, the resulting breeze may be much stronger.

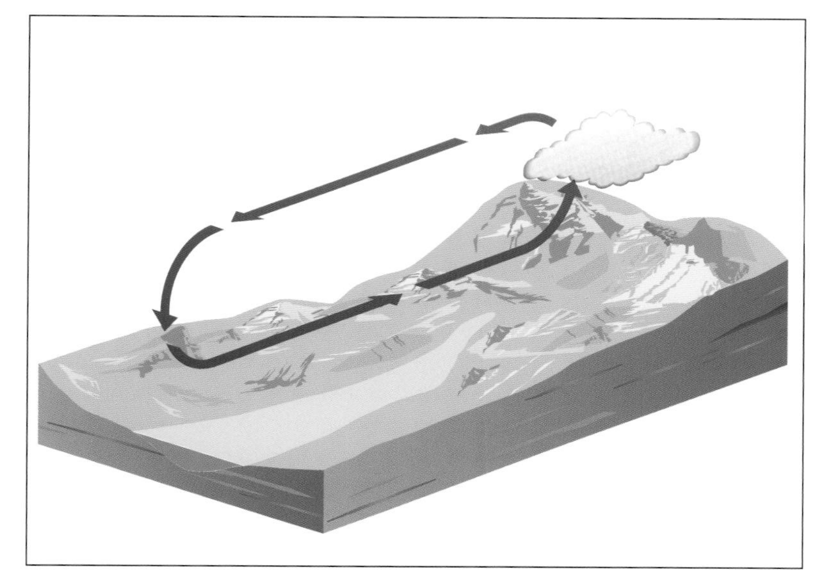

Valley and mountain winds

Heating of mountain slopes during the day gives rise to a flow of air both towards the head of the valley and up the slopes towards the ridges. If fog forms during the night, it will often lift early in the morning, initially becoming a layer of Stratus cloud which later breaks up into individual patches of cloud that are carried up the mountainsides. Such a valley wind usually begins around sunrise, reaches its greatest strength when heating is at a maximum (usually shortly after midday) and dies away at sunset. Over sun-warmed slopes valley winds may reach around 20kph, but over

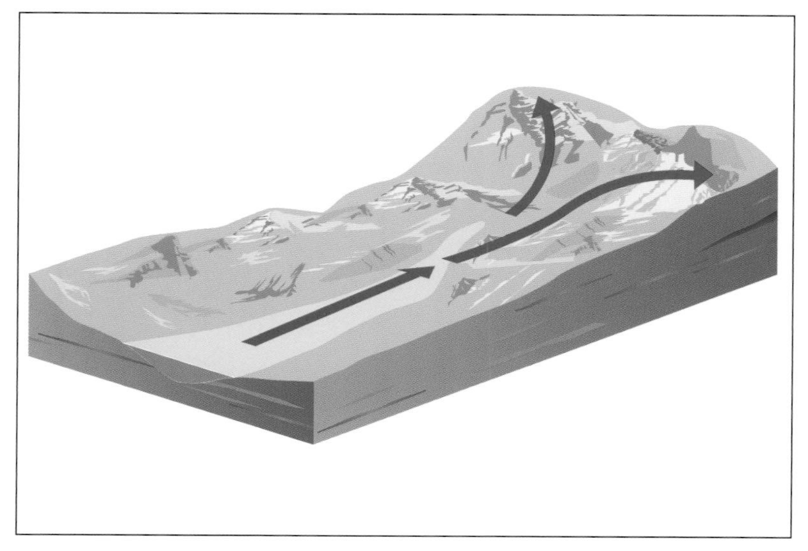

RIGHT A valley wind begins to flow when heating warms the sides and head of a valley. Cloud may appear above the ridges and peaks. *(Ian Moores)*

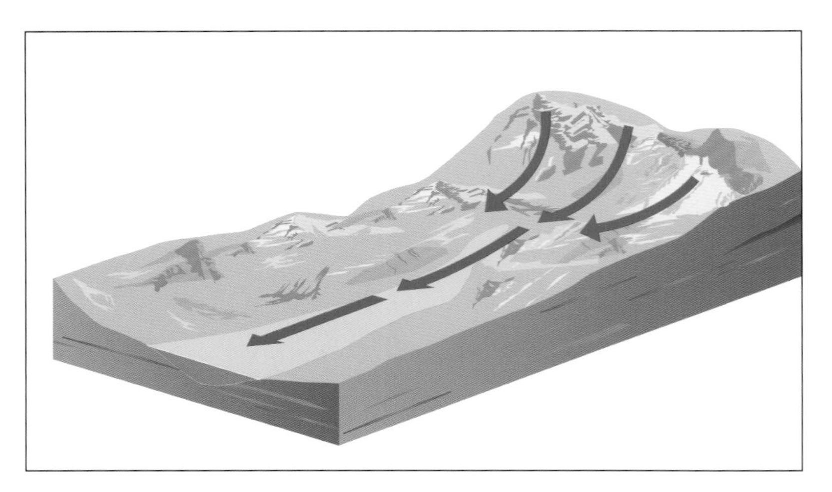

ABOVE When the slopes of a valley cool at night, a downslope mountain wind begins to flow, usually stronger than the corresponding valley wind. *(Ian Moores)*

shaded slopes the wind may be so weak as to be undetectable.

The occurrence of valley winds is greatly influenced by the overall gradient wind, with the strongest winds occurring during warm, anticyclonic conditions. If there is a considerable gradient wind, the resulting turbulence created by the ridges and peaks in hill or mountain regions may prevent valley winds from forming.

A mountain wind is the night-time counterpart of a valley wind. When heating ceases around sunset, the higher ground tends to lose heat more rapidly than the sheltered valleys. The cooled, denser air then slides downwards into the valleys. Such mountain winds tend to be somewhat weaker than the corresponding valley winds, but often reach speeds of 12kph, which may be considerably increased if a valley narrows into a confined gorge or canyon. Mountain winds tend to continue for some time after sunrise, and indeed, in the early morning mist and fog may frequently be seen drifting down the valley, carried by the last of the mountain wind.

More extreme winds, closely related to mountain winds, occur when a pool of cold air builds up over high ground (especially when this is covered in snow or ice). The air in contact with the surface becomes extremely cold and dense and rushes downslope, in what is known as a katabatic wind. Such winds generally affect a much larger area than the localised mountain winds. Many local winds with specific names are of this type, one of the most famous being the fierce, cold bora that affects the eastern coast of the Adriatic (primarily the coast of Croatia). The mistral (already mentioned) is another similar wind. The strongest katabatic winds are found (not surprisingly) around Antarctica, where Commonwealth Bay on the George V coast holds the world record (67kph) for the highest wind speed averaged over the year. (The highest speed ever recorded at the same site is 320kph.)

BELOW Cooling of the air above a snow- or ice-field starts a katabatic wind, which remains cold, despite adiabatic warming as it descends. *(Ian Moores)*

Föhn winds

The temperature of any downslope wind will tend to increase with adiabatic warming, but the extreme examples of this come with föhn winds. As we have seen, when air ascends over a mountain it cools at the dry adiabatic lapse rate (DALR) until condensation sets in, when it cools at the slower saturated adiabatic lapse rate (SALR). On descending to leeward of the hills or mountains the rate at which it warms will depend on whether it has lost any moisture as precipitation. If none has been lost, then it will warm initially at the opposite of the SALR and then, once all the condensation droplets (*ie* cloud particles) have evaporated, inversely at the DALR. If any form of precipitation has been deposited on the hills, however, there will be less water to evaporate, and it will start to warm sooner. It will be warmer at any given level on the leeward side than to windward. Not only is there a 'rain shadow' to leeward of the hills, but temperatures will be warmer. This warming is known as the föhn effect, and may be very considerable. Southerly airstreams from the Mediterranean often deposit rain and snow on the windward side of the Alps, but then warm greatly as they plunge down the northern side of the range.

The sudden onset of föhn conditions may cause a dramatic rise in temperature, which may, for example, swiftly disperse snow cover and, if prolonged, may cause a fire hazard because of the desiccating effect on trees, shrubs and wooden structures. Föhn winds are often accompanied by characteristic long streaks of Stratus or Altostratus clouds that run parallel to the mountain ridge, and known in German as 'föhn Fische' (föhn fish) or, in the case of the chinook, as the 'chinook arch'. Such clouds often take on dramatic colouring at sunrise and sunset, particularly when the mountains (and the clouds) run in a north–south direction.

LEFT Vibrant colour in a chinook arch. Such extreme, dramatic colours are often seen on clouds over south-western Alberta. *(Wikipedia)*

Local effects

Quite apart from the named and local winds, there are many localised effects that may arise. Where the wind encounters a river valley there is a strong tendency – especially if the valley is fairly deep – for the wind to flow along the valley, following its twists and turns. Everyone is familiar with the way in which the wind speed is increased and becomes gusty between buildings. Similarly, wind speeds may increase considerably where the walls of a gorge or a narrow col cut through a range of mountains and create a funnelling effect. The kosava is just such a ravine wind along the valley of the Danube where it cuts through the Carpathian Mountains. A similar effect occurs at sea where the wind speed increases as it funnels between islands, even where the land is fairly low-lying. When the shorelines are steep, the effect may be enough to turn a light breeze into a strong, gale-force wind. This is, for example, the case in the Mediterranean, where the winds in the strait between Corsica and Sardinia (both of which are mountainous) become so strong as to be hazardous to uninformed sailors. Similar conditions apply in the Straits of Gibraltar, where a westerly wind may double in speed. With an easterly wind, the effect is not so marked, but there is a permanent inflow of surface water from the Atlantic (and a corresponding subsurface outflow), and the effects of 'wind against water' may cause areas of the surface to become extremely choppy (particularly close to Gibraltar itself).

There are a number of other weather effects along the coast. As we have seen, when there is increased friction with the surface both the wind speed and the Coriolis force decrease, and the wind tends to back more strongly across the isobars and flow in towards the centre of the low. This effect will occur when winds strike a low-lying coast, but there are other consequences. As the wind slows over the land, there is a tendency for the air to build up just offshore, increasing the strength of the wind. If the shoreline is steep, it may act somewhat like a cold front, causing extra lift, which, under convective conditions, may cause cumuliform cloud to arise along the coast and may, if showers are already present, cause these to become fully fledged thunderstorms. The alteration in conditions occurs along a zone offshore that may be some 5–10km wide. On the landward side, the increased convective activity may extend as much as 50km inland. In the summer this often results in thundery activity during the night.

Even when the wind is blowing along a coastline, there is still a tendency for the wind speed to increase, and the steeper the land, the greater this effect will be. Naturally,

ABOVE If orographic lifting causes precipitation to fall on the windward side of high ground, the air becomes drier and warms at a faster rate as it descends to leeward, creating the föhn effect. *(Ian Moores)*

Cross references

Lapse rates p.51
Wave clouds p.84–87
Wind speeds p.126

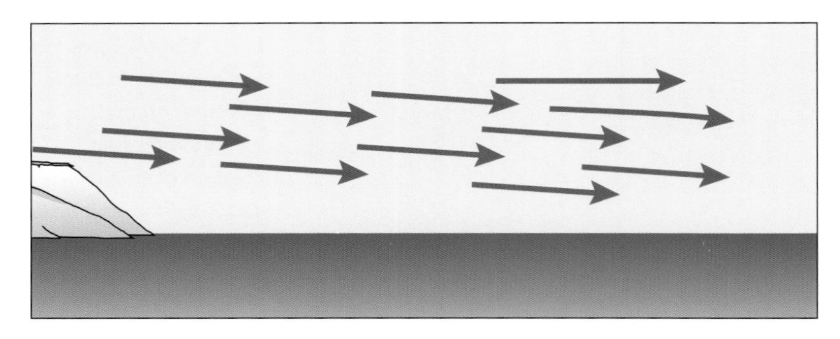

LEFT When conditions are stable, a calm patch extends about 10 times the height of the cliffs out towards the sea. *(Ian Moores)*

when the wind is offshore there are different consequences. There then tends to be a zone of weaker winds, extending out to sea. Again, the steeper the coast, the wider the zone that is affected, and its width may amount to some ten times the height of the cliffs or hills along the shoreline. If there are steep cliffs, air flowing over them may give rise to unpredictable vertical gusts, and under some conditions a major vertical eddy (a pillow eddy) may develop downwind over the sea, where the surface wind is not only gusty, but reverses direction and blows towards the cliff. Conversely, where the wind is blowing on to a steep cliff-face an eddy may develop on the upwind side, over the sea. This is a pillow eddy, and again the surface wind direction may be reversed, thus blowing away from the cliff. Naturally, such effects also occur inland whenever there are steep cliff-faces and mountainous terrain. Such pillow and bolster eddies may be major hazards to aircraft taking off or landing on airstrips within narrow, steep-sided valleys.

When the wind is roughly parallel to the coast, any cape or headland will create its own effect on the wind. If the land is fairly low it will not impede all of the flow, but there will be a tendency for the wind to strengthen just offshore and to curl round behind the headland. On the side that is downwind to the gradient wind there may even be a tendency for an opposing flow to be created towards the projecting cape. When the shoreline is very steep these effects will be amplified to such an extent that a major horizontal eddy may be created behind the cape, and even a small, localised low-pressure area. Similar effects occur, of course, in mountainous areas on land, and low-pressure areas behind isolated peaks are believed to contribute to the formation of banner clouds.

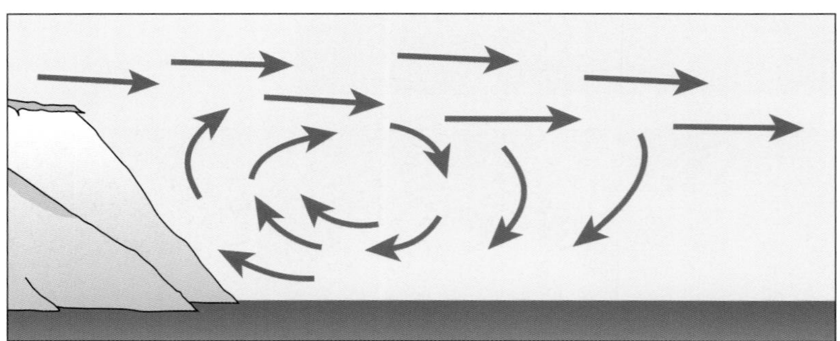

ABOVE When conditions are unstable (with cumulus clouds, for example), a pillow eddy tends to form below the cliffs, with a gusty and fluky wind. *(Ian Moores)*

BELOW When the wind is onshore, steep cliffs tend to create a bolster eddy to windward, again with a gusty wind. *(Ian Moores)*

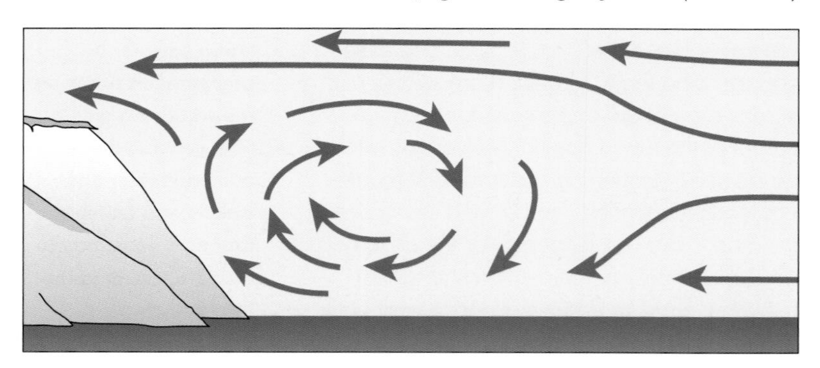

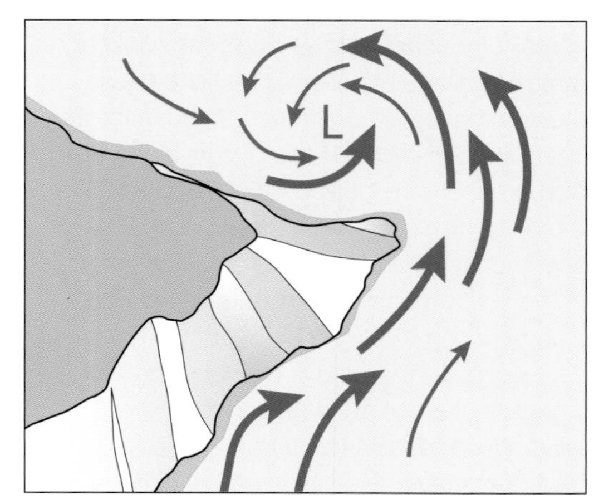

LEFT When the wind is along-shore, a pronounced low-pressure area and eddy may develop behind a cape with high ground. *(Ian Moores)*

Cross references

Banner clouds p.86–87
Cold front p.43
Coriolis effect p.19

Chapter 10

Showers and storms

Showers

To most people the word 'shower' simply means a short period of rain, but to meteorologists it has a specific meaning: convective rain, *ie* rain from Cumulus congestus or Cumulonimbus clouds that falls over a relatively limited area, rather than the rain from frontal clouds (Nimbostratus) that may affect an extremely extensive area of the surface. Occasionally, as already mentioned, convective clouds may be embedded in frontal cloud, especially in cold fronts, giving a localised patch of very heavy precipitation. Powerful Cumulonimbus clouds may, of course, give rise to hail or develop into fully-fledged thunderstorms.

Both Cumulus congestus and Cumulonimbus clouds occur only when there is extensive instability and convection. Some indication of the extent of convection is given by the amount of sky that is covered by clouds (where air is rising) relative to the clear areas, where air is sinking to balance the air rising in thermals. As a very rough guide, therefore, showers are likely when the area covered by clouds exceeds the area of clear blue sky. In addition, shower clouds are usually very deep and have very dark bases.

The duration and area affected by showers and the intensity and form of precipitation is strongly dependent on the time of year. In winter, in particular, the freezing level is relatively low. Precipitation is initiated by the Bergeron-Findeisen (freezing) process. However, the clouds are shallow and have a limited water content and precipitation will be initiated as small ice or snow pellets or snowflakes. Depending on the temperature profile these may melt into rain before they reach the surface.

In summer, temperatures are higher, clouds are deeper and their water content is much greater. The freezing level is also at a greater altitude. During a cloud's early lifetime, at the

LEFT Rain began falling from the dark base of this heavy Cumulonimbus cloud shortly after the photograph was taken. Rain is already falling from another cell in the distance. *(Author)*

BELOW A very shallow winter-time Cumulonimbus, producing heavy precipitation over the sea, in a very cold north-westerly airstream. *(Author)*

RIGHT An angry-looking massive Cumulus congestus cloud, just about to become glaciated, seen from the air. The aircraft landed ahead of the shower, which had developed to the fully mature stage in about 30 minutes. Passengers were unable to leave the terminal because of torrential rain and large hailstones that lasted about 10 minutes. *(Author)*

Cumulus congestus stage, raindrops may form through the 'warm' process, that is by collision and coalescence. Later, when the clouds have become Cumulonimbus, glaciation will occur and the 'cold' process will come into play. Within any shower cloud the amount of water in raindrops is, initially, fairly small, and most droplets will be suspended in the vigorous updraught. As the amount of water in raindrops increases and they grow in size, they weaken the updraught, which eventually collapses, releasing the raindrops to fall as a heavy shower.

When a Cumulonimbus cloud is particularly deep, reaching well above the freezing level, initially the cloud droplets tend to become supercooled, but at about -10°C any freezing nuclei present start to act and some droplets freeze into tiny ice crystals. If the cloud is even deeper, and the temperature drops to -40°C, the supercooled droplets freeze spontaneously, and the cloud has become fully glaciated. Any liquid droplets that collide with ice crystals freeze rapidly to produce particles of hail. Small particles may be swept upwards through regions at different temperatures and (as we have seen) grow in layers until they become too heavy to be supported by the rising air, and fall out as hail.

BELOW A column of heavy rain from the mature stage of a single Cumulonimbus cell. Eventually the downdraught will quench the updraught of warm, humid air feeding the cloud, and the rain will cease. *(Duncan Waldron)*

There are three recognised stages in the development and decay of individual Cumulus congestus and Cumulonimbus clouds:

- Early (growing) stage: Cumulus congestus.
- Mature stage: Cumulonimbus calvus, followed by Cumulonimbus capillatus, giving the greatest precipitation.
- Late (decaying) stage: gradual decay and lessening precipitation.

The duration of each of the first two stages is about 20 minutes. There may be some large raindrops during the early stage, but most of the precipitation will occur in the mature stage, initially as large raindrops but then possibly as heavy rain accompanied by hail. The decaying stage occurs because the powerful downdraughts overwhelm the updraughts, cutting off the supply of warm humid air to the cell. This stage may last from about 30 minutes to as long as two hours, during which time the rain lessens and the individual drops become smaller. In general, a single Cumulonimbus cell has a lifetime of about 90 minutes.

Small, individual cells may cover an area of 10–12km^2, and rainfall may last between 10 and 30 minutes. Over land, daytime heating is largely responsible for the convection, so showers tend to die away at night. When the activity arises because cool air is passing over a warm sea, convection may continue to create shower clouds throughout the night.

If the gradient wind is light, the whole sequence may be observed in individual isolated cells. With stronger winds, there is a tendency for several cells to be present in various stages of development. If there is a strong wind shear, the top may precede the main body of the cloud and produce some precipitation ahead of the heaviest rain. The duration of convective activity and rainfall tends to increase.

With strong convection, the rising cells may reach a stable layer (which may even be the tropopause) and spread out beneath it. Unlike the Stratocumulus or Altocumulus (Sc cugen and Ac cugen) that arise when normal Cumulus clouds encounter an inversion, this cloud tends to resemble Altostratus or Nimbostratus (As cugen or Ns cugen). The

cloud is usually distinctly fibrous in appearance and is accompanied by large amounts of Cirrus fibratus. With strong wind shear, a large anvil (Cb incus) may be created, often with mamma hanging beneath it.

When the gradient wind is strong, showers tend to be short-lived and there are fairly long periods of clear skies and sunshine between them. When the wind is weak, showers tend to be of longer duration, and there are therefore less of them during a day. Any Cumulonimbus cloud tends to create blustery conditions around it, because of the vigorous updraughts and downdraughts within it. The updraughts draw air in towards the cloud from the surroundings and may thus create a 'wind' that seems to blow towards the advancing cloud. This may cause confusion in inexperienced observers and is the cause of the statement, sometimes heard, that a shower approached 'against the wind'. The downdraughts, in contrast, may produce powerful gusts that spread out ahead of the cloud. The changes in the direction and strength of the surface wind at such a gust front are common and may be very considerable. Air flowing into the cloud is often undercut by a cold downdraught, giving rise to a distinct roll or shelf cloud on the leading edge.

The air drawn in towards a Cumulonimbus cell will often initiate the formation of an additional cell, so that eventually there may be a cluster of cells, at different stages of growth and evolution. Such secondary cells prolong the life of the overall cluster, but because they normally form to one side of the initial cell they bring rain to different areas of the surface.

Showers usually cause a drop in surface temperature, often of about 3°C, because of the cooling effect as the precipitation evaporates in the warmer air beneath the cloud.

FAR LEFT The rear of the anvil of a Cumulonimbus capillatus cloud that was just reaching the mature stage and about to give rise to precipitation. The striated and fibrous nature of the cloud-top is clearly visible. *(Author)*

LEFT Individual Cumulonimbus cells, one of which has reached the anvil stage, created by cool air flowing over the relatively warm sea. *(Author)*

LEFT Roll and shelf clouds, created by strong up- and down-draughts ahead of an advancing, very active Cumulonimbus cloud. *(Author)*

Cross references

Bergeron-Findeisen process p.96
Glaciation p.53
Hail p.99
Mamma p.76
Supercooling p.166
Wind shear p.166

LEFT This large, active Cumulonimbus cell is drawing air into the system on its rear flank (left), giving rise to a series of young cells in various stages of growth, which later produced considerable precipitation and developed into a thunderstorm. *(Author)*

Thunderstorms

Large, active Cumulonimbus clouds frequently develop into thunderstorms. Despite years of research, there is still no generally accepted theory that accurately accounts for the way in which electrical charges become separated within a cloud, nor what precisely initiates the discharge. It is known that cloud temperatures must be below -20°C, and that water droplets and ice crystals must both be present in the cloud. The creation and separation of the charges appears to be related to the freezing and fragmentation of ice particles, high in the cloud. The lighter, positively charged particles are carried to the top of the cloud, and the heavier, negatively charged particles accumulate at the cloud's base. As the cloud is carried across the surface by the wind, the charged base induces a positive charge on the ground beneath it. This positive charge migrates across the ground underneath the cloud until, eventually, either the charge becomes so great or the distance between the base of the cloud and the ground becomes so small – normally above a tall object such as a high building or a tall tree – that the electrical resistance of the air breaks down, and a lightning discharge occurs.

However, such a mechanism does not account for all lightning discharges, which may occur within the cloud itself (intracloud lightning), between nearby clouds (intercloud lightning), as well as to the ground. Various unconfirmed suggestions have been made for the trigger for such discharges, including the electrical breakdown being initiated by cosmic rays from space. Many different electrical phenomena that accompany thunderstorms have been discovered in recent years, as well as the fact that lightning discharges are associated with both X-ray and high-energy gamma-ray discharges, and even the presence of positrons (the antiparticle counterpart to the electron), but the precise relationships between all these phenomena and the mechanisms involved remain obscure.

BELOW **A lightning strike to the sea at dusk, clearly showing the branched structure, typical of 'fork' lightning, together with the main channel through which the main discharge occurred. A strike from an unrelated active cell is visible on the left.** *(Duncan Waldron)*

A lightning strike is extremely hot – hotter than the temperature of the surface of the Sun – and the channel of air heated by it expands and then collapses at supersonic speeds, producing the sound of thunder. The light from the flash reaches the observer almost immediately, of course, but the sound travels much slower, so that counting the seconds between seeing the flash and hearing the thunder gives an indication of the distance of the lightning strike: about three seconds per kilometre (five seconds per mile). If the bearing of a sequence of flashes remains constant, then the cell generating the flashes is likely to pass over you. However, as with the rainfall from Cumulonimbus, the lifetime of the cell causing the electrical activity is limited to 20–30 minutes, so activity may cease before the cell passes overhead. If flashes are seen, but no sound is heard, the cell is probably at least 25–30km distant.

People sometimes differentiate between 'fork lightning' (when the actual lightning channel is seen) and 'sheet lightning' (when it is invisible), but in fact there is no difference in the processes at play. So-called 'sheet lightning' occurs when the discharge occurs within a cloud, or between two clouds and the lightning channel is hidden by intervening cloud. Another common misconception is that there is something special about 'heat lightning', which is thought to occur in summer. In fact this is merely distant lightning, often beyond the horizon, that is occurring so far away that no sound is heard. Lightning may sometimes strike many kilometres away from the parent cloud, seemingly from a clear sky well outside the area covered by the cloud. The existence of such 'bolts from the blue' is well established, and they are one reason why care should be exercised whenever thunderstorm activity is present.

ABOVE LEFT Four separate lightning discharges from two separate, but neighbouring, cells. Only the most distant dischage shows the characteristic branching structure. *(Duncan Waldron)*

ABOVE Cumulonimbus clouds illuminated from inside by 'sheet' lightning. One cloud clearly shows how lightning may jump from one part of the cloud to another. *(Duncan Waldron)*

A lightning discharge takes place in several stages. In general, a stepped leader first makes its way from the cloud down towards the ground. This usually has a branched structure where several individual paths are initiated. When one of these reaches the ground or a high structure, a discharge channel is established and the main current normally travels upwards from the ground into the cloud. There may be a sequence of return strokes between cloud and ground. Usually these occur too rapidly to be apparent to any observer, but occasionally they may follow one another more slowly, and give a flickering appearance to the stroke. Such multiple channels may be photographed by special cameras, but are, on rare occasions, visible to the naked eye if the observer is moving at a fairly high speed – on a train or in an aircraft, for example – roughly at right angles to the line of sight to the discharge. Occasionally the positive charge at the top of the cloud becomes so great that a discharge occurs between the cloud-top and the ground and the main discharge current flows in the opposite direction to that found in the majority of strikes. Such 'positive' discharges generally carry an even greater current than the normal ground-to-cloud discharge currents.

Multicell storms

The action of drawing air into an active cell frequently initiates the formation of additional cells, which may form a 'flanking line' outside the active centre. These new cells may thus give rise to a cluster of cells that are active simultaneously, producing a multicell storm. From a distance it may be possible to determine the individual cells that make up the cluster.

Sometimes the activity is limited to heavy rain and hail, but frequently such clusters also exhibit lightning activity. Although each individual cell may be active for 30–60 minutes, such multicell storms may be extremely persistent, lasting several hours as new cells are generated, and travel for long distances across country. If lightning is present, the location of

BELOW Air drawn into an existing cell tends to create a new cell where the cold downdraught on the gust front lifts warm, humid air away from the surface. *(Ian Moores)*

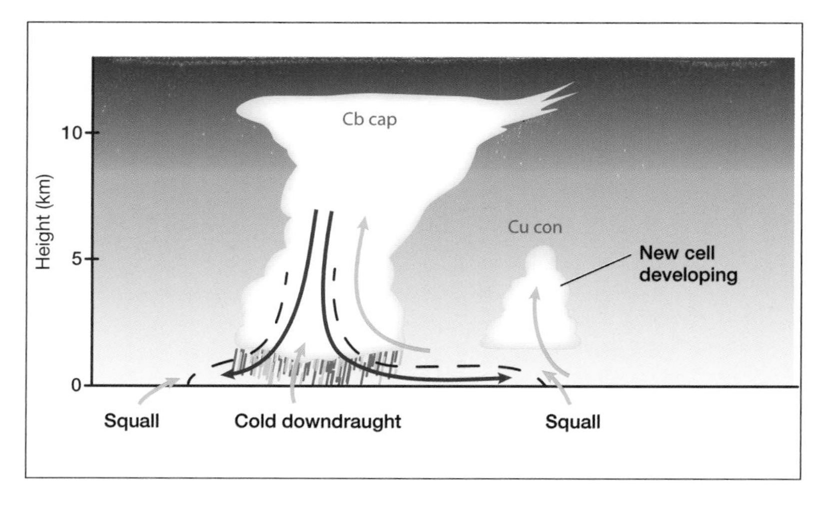

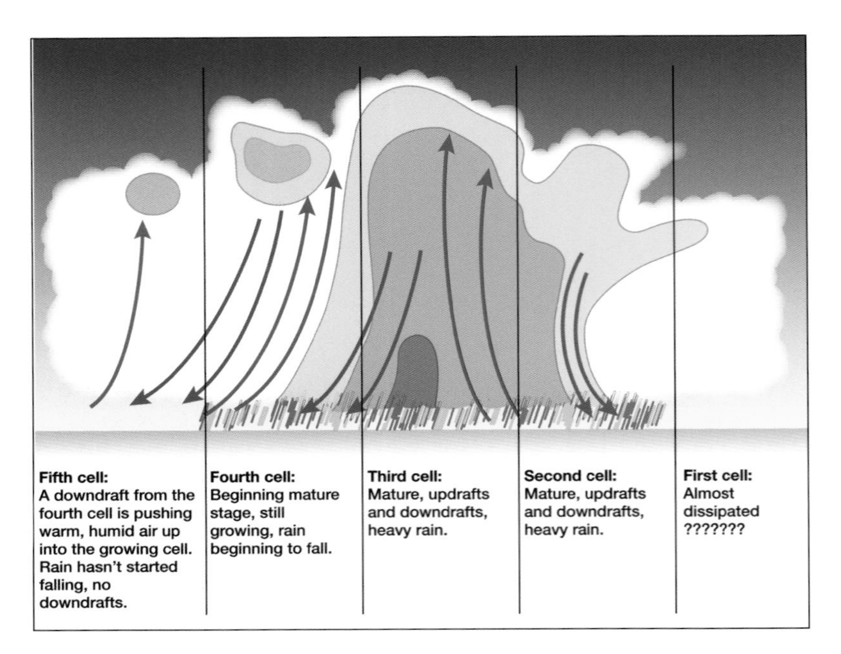

ABOVE **A multicell cluster, here consisting of five cells. The oldest cell on the right has nearly dissipated. Most of the rain is being produced by the third, mature cell.** *(Ian Moores)*

individual lightning strokes reveals the active cells, enabling one to determine whether the wind will carry an active cell overhead. Activity may die away from the oldest cell, but continue in other cells within the overall system.

Supercells

An even more extreme form of storm is known as a supercell. These generally arise when there is a very deep pool of unstable air and the wind speed increases and veers considerably with height. Instead of the activity being organised into individual cells, a single system arises, consisting of a giant rotating column of rising air, known as a mesocyclone. The top of the cloud may be between 8 and 15km high. A system is accompanied by a complex pattern of updraughts and downdraughts, with cool air entering the system at middle levels. In normal Cumulonimbus cells the downdraughts tend to suppress the flow of rising air, leading to a cell's limited lifetime, but in a supercell the updraught is tilted and the whole system tends to move rapidly across the countryside. The cold downdraughts are unable to quench the updraughts, so the system is very persistent and often lasts for six hours or more. The rotation of the overall mesocyclone tends to keep the various updraughts and downdraughts separate from one another, adding to the system's persistence. Supercells generated over France in summer, for example, sometimes cross the English Channel, bringing severe storms to southern England, and may penetrate far inland.

The circulation within a supercell tends to be such that a giant 'vault' is created where the updraught is strongest. Such a situation is ideal for the formation of large hailstones, which may be carried upwards several times by the strong updraughts, gradually accreting further layers of ice until they become too heavy to be sustained and fall out of the cloud.

Supercells are the most powerful storms and frequently produce torrential rain; large, damaging hail; and multiple lightning strokes. The downdraughts themselves may be so powerful that they cause damage on the ground, but, more significantly, supercells often create highly destructive tornadoes. Supercell storms are generally most frequent at middle latitudes in summer, especially over the central and eastern states of the USA, as well as other continental areas around the world. They may occur at any time of year in the tropics.

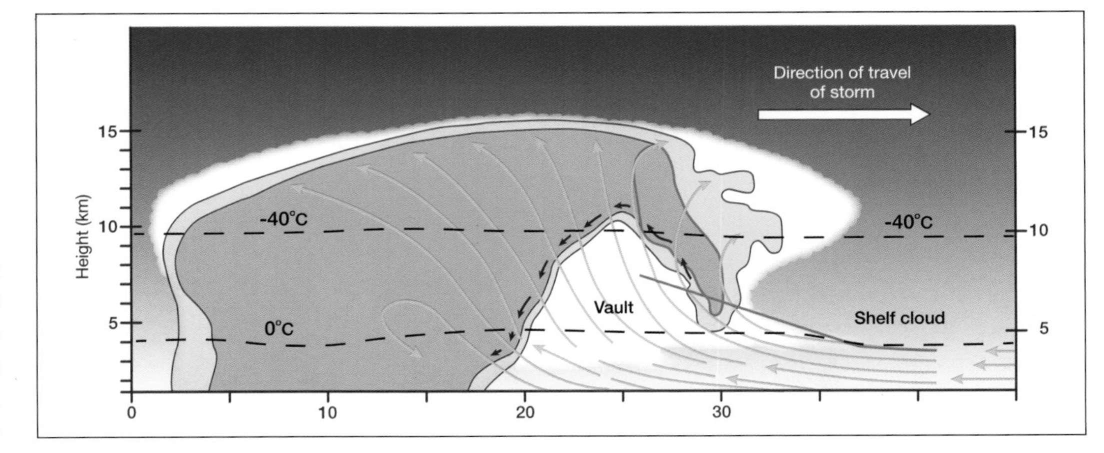

RIGHT **Buried within a supercell is a large rain- and hail-free vault produced by the vigorous updraught.** *(Ian Moores)*

ABOVE The overall rotation of a supercell mesocyclone is often clearly visible in the structure of the clouds as in this storm illuminated by the setting Sun over New South Wales. *(Duncan Waldron)*

Mesoscale convective systems

Frequently, Cumulonimbus clouds may become organised into a much larger cluster or a line that exhibits major instability. The clusters of convective clouds, known as mesoscale convective systems (MCS), may consist of multiple Cumulus congestus or Cumulonimbus, together with associated stratiform clouds. They exhibit a vigorous circulation and heavy precipitation. The circulation becomes stronger as time passes, leading to a deep layer of cold air in the lower troposphere. The tops of the clouds tend to combine to produce a giant anvil or Cirrus shield. Such systems persist for four hours or more – much longer than individual Cumulonimbus cells.

LEFT Two prominent thunderstorm clusters are visible over northern France, together with a mesoscale convective system that covered practically the whole of Belgium, in this infrared image obtained on 25 May 2009. *(Eumetsat)*

LEFT A mesoscale convective system covering most of the Czech Republic on 13 July 2011. Formed shortly after sunset over Bavaria and the Czech Republic, it moved to south-western Poland where it dissipated before local noon. *(Eumetsat)*

LEFT A satellite image showing the development of lines of Cumulonimbus, Cumulonimbus clusters and supercells. There is a large supercell in Texas and an enormous supercell system covering parts of Texas, the Oklahoma Panhandle, and a portion of Kansas. *(Author)*

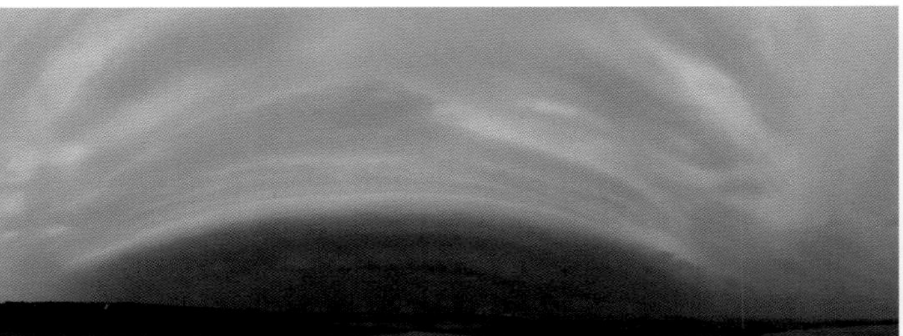

BELOW Three photographs of the mesoscale convective system of 13 July 2011, obtained when the system was over southern Bohemia (Czech Republic). *(Eumetsat)*

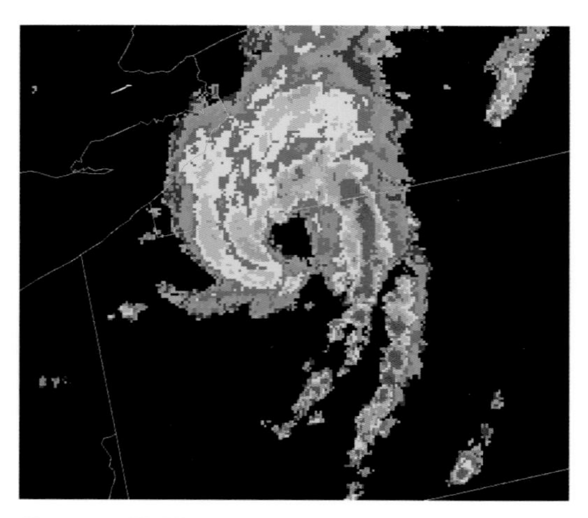

RIGHT This satellite image shows a mesoscale convective system, just to the east of Lake Erie on 21 July 2003, which actually resulted from the combination of the 'hook' region of a supercell and a long, trailing squall line. *(Wikimedia Commons)*

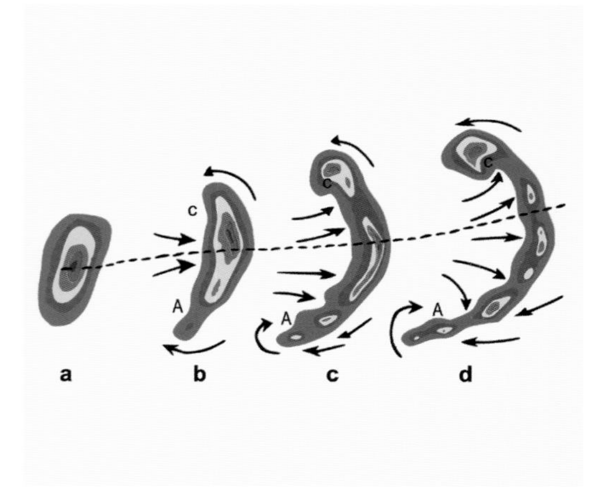

FAR RIGHT The development of a thunderstorm radar echo from an initial concentrated cluster into a bow echo (b, c) and then into a comma echo (d). The regions (A & C) at the ends of the echo, exhibit rotation and (especially C) are often the site of tornado formation. *(Wikipedia Commons/Utahweather)*

Squall lines and downbursts

If a mesoscale convective system assumes a linear or curved form, it is known as a squall line. This may either be part of a cold front or a line ahead of a normal cold front, and be hundreds of kilometres long and sweep across country. New cells are generated by the vigorous outflow ahead of the squall line. The squall line itself may be so strong that there is a slight rise in pressure where precipitation is greatest, and a drop in pressure behind the line.

Radar echoes reveal how the initial concentration of activity spreads out into a curved shape – in what is known as a bow echo – and then later progresses into a comma echo. The region in the centre, where the curvature is greatest, is often the site of extreme downdraughts – known as downbursts – which may strike the surface with great intensity and spread out in a radial pattern. Such downbursts are a major hazard to aircraft, especially when they are taking off or landing. Downbursts may also cause extensive damage on the ground.

An extreme form of squall line may be 200km long (or more) and is known as a derecho. These may be highly destructive because hurricane-force winds may be experienced behind the initial gust front. (In derechos the wind speed increases, rather than decreases, behind the leading edge of the gust front.)

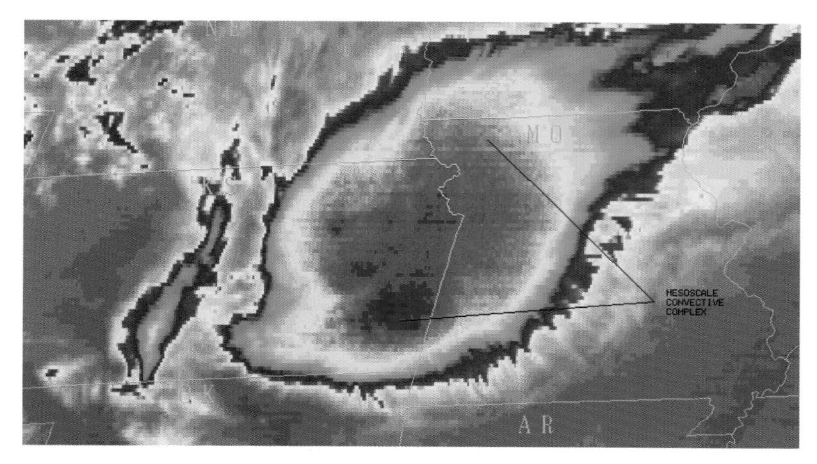

BELOW A mesoscale convective complex (MCC) forming over Kansas and western Missouri on 8 July 1997 at 11:45 UT in an infrared image from the GOES-8 geostationary satellite. *(NASA)*

Mesoscale convective complexes

Sometimes particularly large and vigorous mesoscale convective systems combine into a giant convective system, known as a mesoscale convective complex (MCC). These systems are actually defined by their properties as determined from infrared satellite observations. There must be an area greater than 50,000km^2 with a temperature below -52°C, and a larger, elongated and continuous area of at least 100,000km^2 with a temperature below -32°C. These conditions must persist for at least six hours for a system to be classed as an MCC. These systems tend to grow late in the day or at night as smaller clusters of cells merge into an organised system, and often persist into the next day. Such gigantic organised systems are found over the Americas, Africa and Asia. When such a system travels from the land out over the sea, it may become the seed for the formation of a tropical cyclone.

Chapter 11

Uncommon and severe weather

Whirls

There are a number of events that may be classified under the general heading of 'whirls'. Some of these (such as tornadoes) may, of course, be extremely violent and destructive, but many are fairly weak, although all tend to be associated with some form of disturbance to the surface. Whirls include:

- Devils.
- Funnel clouds (tuba).
- Landspouts.
- Waterspouts.
- Gustnadoes.
- Tornadoes.

Devils

The general class of whirls known as devils may form in two distinct ways. Perhaps the most common are whirls that are simply set in motion by the wind. Whirls of leaves, paper and other litter are often created when the wind is funnelled by the buildings (particularly tall buildings) in towns and cities. The wind acting on the sides of a valley or cliffs may create similar whirls, which, if they travel across a lake, for example, may produce a water devil. Other surfaces may give rise to snow devils, dust devils or even hay devils.

A different mechanism comes into play when there is intense heating of the ground. Strong convection may arise, and any inherent rotation may be amplified if the surface has areas of differing roughness. Adjacent areas of smooth and rough terrain may increase the tendency of the rising air to rotate and lift material from the surface. Dust devils commonly arise in hot, arid areas through these mechanisms – they have even been observed frequently on Mars. On rare occasions, the rising air may reach a sufficiently high altitude for condensation to occur, producing a small Cumulus cloud above the devil. Although such dust devils may travel across the countryside, they are rarely strong enough to create major damage.

Waterspouts and landspouts

Strong convection within cumuliform clouds may produce a rotating column of air that descends towards the surface. This commonly appears as a grey funnel cloud (or tuba) and, in the majority of cases, does not reach the surface. When it does, however, the funnel cloud becomes converted into a landspout if over the land, or a waterspout if over the sea. Such whirls are not particularly uncommon, and may sometimes occur when the sky appears to be covered in Stratocumulus clouds and one would assume that little strong convection is present. However, individual cumuliform clouds

BELOW This well-developed dust devil was photographed in Nigeria. *(Ron Livesey)*

RIGHT The early stages of the development of a waterspout, when a funnel cloud (tuba) starts to descend towards the surface. *(Mike Spenkman)*

FAR RIGHT The funnel has now reached the sea surface and a cloud of spray (the bush) has now become visible. *(Mike Spenkman)*

(Cumulus congestus or even Cumulonimbus) and their associated strong downdraughts may be present, but masked by the general layer of cloud. The temperature of the surface also plays a part, particularly in the formation of waterspouts. Conditions are especially favourable when the sea surface temperature is warmer than the overlying air, such as cold polar air arrives behind a cold front, creating an unstable situation that favours downdraughts from the cloud and updraughts from the warm surface.

RIGHT A waterspout that formed over the Gulf of Mexico. The 'bush' of spray is clearly visible, and there are indications of the hollow centre within the condensation funnel. *(NOAA)*

Both landspouts and waterspouts have a common structure. There is a central rotating core, consisting of a strong downdraught, which becomes visible as the funnel cloud. Outside this there is a rotating updraught, but this is normally invisible. There is a spiral inflow at the actual surface, and a relatively small amount of material may be lifted into the air. In the case of waterspouts, the point where the spout's downdraught first touches the water is known as the 'dark spot' and there is a ring of spray thrown out at the base. When the wind speed rises beyond a value of about 80kph a distinct cylinder of spray appears, which is known as the 'bush'. This superficially resembles the debris cloud raised by a far more destructive tornado.

Waterspouts (and landspouts) are relatively short-lived, with lifetimes that average 15 minutes, and have diameters of 15–30m. Waterspouts often dissipate if they cross on to land, although they may penetrate a short distance inland. The longest recorded waterspout had a height of about 1,000m, but the diameters and sizes are very variable. Multiple funnel clouds and waterspouts are frequently observed. (Multiple landspouts probably occur with equal frequency, but are more difficult to observe because of the many obstructions to the line of sight that exist over land.)

Gustnadoes

Phenomena that are essentially similar to landspouts and waterspouts are often generated by the strong winds and convection that occur at gust fronts and the intense activity associated with tropical cyclones (hurricanes). In general, these whirls do not create a condensation funnel and are not particularly strong. Some of the damage and destruction reported as being caused by the storm may actually have been created by these relatively short-lived whirls. It is these gustnadoes and landspouts (in particular) that are often described as 'twisters' by reporters for the various media.

Tornadoes

Tornadoes are in a class of their own, and arise through a completely different mechanism from that found in the weaker waterspouts, landspouts and gustnadoes. Tornadoes are uniquely associated with the vast rotating column of air (the mesocyclone) found in supercell storms. The exact mechanisms involved are complex and still not fully understood, but it is known that wind shear creates a cylinder of air that initially is rotating about a horizontal axis. The strong updraughts associated with a growing Cumulonimbus cloud raise this cylinder towards the vertical, creating a U-shaped loop with a pair of vertical vortices, rotating in opposite directions.

For complex reasons, the vortex with clockwise rotation dies away, and the one with anticlockwise rotation persists and strengthens. Air flowing into the storm at middle levels and humid air from the surface create a mesocyclone, a giant column of rotating and rising air that may be between approximately 2km and 20km in diameter, which, however, does not reach down to the ground. The humid air rising into the storm at the base condenses into the distinctive circular cloud, known as a wall cloud. The rotation within the mesocyclone accentuates the much smaller, strong, rotating updraught to such a degree that it may extend down towards the ground from within the wall cloud. The pressure-drop in the centre of the narrow column of rising air is reliably estimated to be 200–250hPa, causing immediate condensation of the humid air into the tornado funnel itself. Once this reaches the ground the system has become a fully fledged tornado, characterised by a debris cloud surrounding the main updraught.

The meteorological conditions that favour the formation of tornadoes are particularly common in the United States, but major tornadoes occur in many parts of the world. One area where tornadoes are frequent, but poorly studied (so accurate statistics and other information is hard to obtain), lies at the head of the Bay of Bengal, in Bangladesh. The very greatest death toll that is regarded as reasonably reliable occurred around Shaturia, Bangladesh, on 26 April 1989, when recorded deaths were in the region of 5,000, and over 50,000 were made homeless.

Tornadoes may be extremely destructive, and their strength is defined on the Enhanced Fujita Scale (a development of the original Fujita Scale, which had certain disadvantages).

It should be noted that this scale is based on estimated wind speeds (in mph), rather than actual measurements, together with three-second gust speeds, and an assessment of damage caused on a 28-point scale. The TORRO Scale, devised by the British Tornado

Cross references

Tropical cyclone p.124
Tuba p.77

BELOW A line of three large supercells, the northernmost of which spawned the devastating EF5 Moore tornado of 20 May 2013. Note the almost complete absence of clouds to the west of the dryline. *(NASA)*

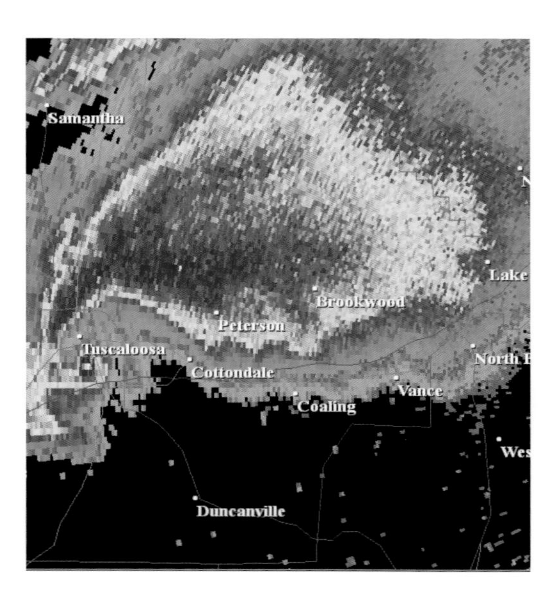

RIGHT A radar reflectivity image of the supercell that produced the high-end EF4 Tuscaloosa tornado on 27 April 2011. The distinctive 'hook echo' that marks the location of the tornado is clearly visible over Tuscaloosa. *(National Weather Service)*

ABOVE RIGHT An apartment block completely destroyed by the Tuscaloosa tornado of 27 April 2011. *(National Weather Service)*

RIGHT The highly destructive tornado that hit the town of Moore in Oklahoma on 20 May 2013. *(Wikimedia Commons)*

and Storm Research Organisation, is based upon actual wind speeds, rather than on assessments of damage. It is thus perhaps more appropriate when accurate measurements of wind speed have been obtained by Doppler radar methods or by other means.

The diameter of mature tornadoes at the surface varies considerably, lying between 100 and 2,000m. One of the very largest occurred during the tornado outbreak of 18–21 May 2013. This was the giant EF5 tornado that hit the town of Moore, Oklahoma, on 20 May, and was 2km across. (There were 24 deaths in Moore itself.)

The majority of tornadoes are short-lived, having lifetimes of about 15 minutes. On many occasions, however, they last for

Cross references

Mesocyclone p.116
*Supercell p116**
Wind shear p.166

RIGHT A photograph giving an overhead view of tornado damage in Moore, Oklahoma on 21 May, 2013. *(Wikimedia Commons)*

RIGHT A composite radar image showing the track of the destructive Tuscaloosa and Birmingham, Alabama tornado of 27 April 2011. *(UCAR)*

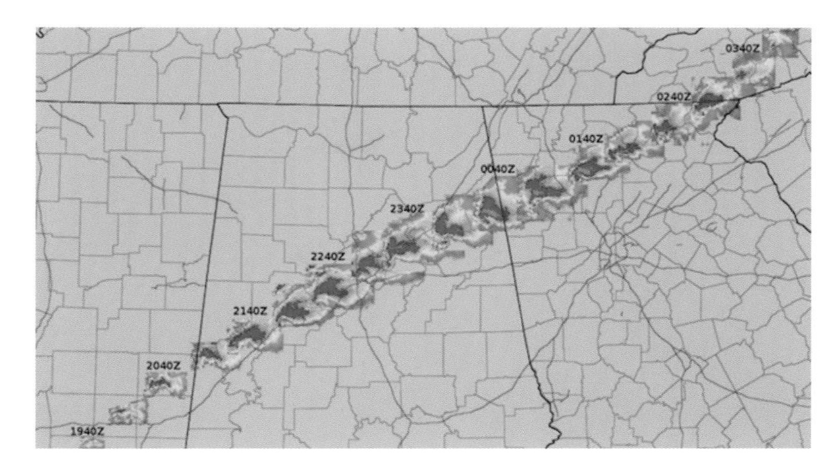

much longer. The supercell that spawned the highly destructive EF4 or EF5 tornadoes over Tuscaloosa and Birmingham (both in Alabama) was tracked by radar for seven hours, during which it covered more than 610km from its origin over Mississippi to its final decay over North Carolina.

Typical path-lengths are 10–100km, but the longest track believed to have occurred was the 472km of the Tri-State tornado of 26 May 1917, although there is a possibility that this was actually the result of a series of individual tornadoes forming and decaying one after another. Multiple outbreaks of tornadoes occur quite frequently, the most known being the 148 events recorded for 3–4 April 1974.

Tornadoes are highly destructive and it is only in recent years, with the use of Doppler radar systems, that accurate measurements have been recorded of the wind speeds that are attained. The highest recorded to date was the 514kph found in the tornado that devastated the outskirts of Oklahoma City on 3 May 1999.

BELOW One of the tornadoes in the outbreak that included the destructive Oklahoma City tornado of 3 May 1999. The base of the dark condensation funnel is surrounded by a partially translucent dust cloud. *(Daphne Zaras/ Wikimedia Commons)*

BELOW LEFT This image was acquired by the Enhanced Thematic Mapper plus (ETM+), aboard Landsat 7, one minute before Earth Observer-1 (which produced an image with the Advanced Land Imager (ALI) instrument passed over La Plata on 1 May, 2002. Although the ETM+ does not have quite the same resolution or sensitivity as the ALI (a maximum or 15 metres per pixel for the ETM+, and a maximum of 10 metres per pixel for the ALI), it has a much wider field of view. The scene shows the entire length of the tornado's path – roughly 39km (24 miles). *(NASA)*

ABOVE The tropical storm Catarina over the South Atlantic on 26 March 2004. Estimated wind speeds at this time were below 100kph, which would mean that they were not strong enough for the system to be classified as a Category 1 hurricane. *(NASA)*

Tropical cyclones

Tropical cyclone is the technical term for the violent, rotating storm systems that have different names in different parts of the world:

- Cyclone: Indian and western Pacific Oceans.
- Hurricane: North Atlantic and eastern Pacific Oceans.
- Typhoon: north-western Pacific Ocean.

Missing from this list of oceans is the South Atlantic. Until 2004 no tropical cyclone had ever been recorded in the region. A major system, named Catarina, developed in March 2004 and caused fairly severe damage on the coast of Brazil. The majority of meteorologists considered it a true tropical cyclone, although the Brazilian meteorological service disagreed, rating it a tropical storm (a system below a tropical cyclone in intensity).

A tropical cyclone is a large (synoptic-scale), non-frontal, low-pressure system with sustained wind speeds in excess of 33m/s. The central pressure is often 950hPa or lower. (The lowest pressure ever recorded, 870hPa, was found in Typhoon Tip over the western Pacific on 12 October 1979.)

Tropical cyclones have a very specific structure. There are bands of very deep convective clouds (*ie* Cumulonimbus clouds) that spiral in towards the centre. Here, at the cyclone's mature stages, there is a cloud-free eye, where air is descending from upper levels. The eye is surrounded by an eyewall in which the most powerful convection, the most intense precipitation and the strongest winds occur. The towering clouds themselves typically reach altitudes of 12km (40,000ft) or more, at which altitude they spread out into a Cirrus shield that may have a diameter of 650km or more. (In Hurricane Gilbert in September 1988, the diameter of the Cirrus shield reached 3,500km.) Tropical cyclones are difficult to study from the ground, because of the extreme and complex conditions occurring within them. With the advent of satellite imagery, structure and development of tropical cyclones may be followed in detail, although the acquisition of data by 'hurricane hunter' aircraft flights across the cyclone remains vital in understanding and trying to predict the behaviour of such systems. Time-lapse photography using satellite images clearly reveals certain features of these systems, such as the way in which air spirals in towards the centre at low levels, and spirals outwards in the overlying shield of Cirrus clouds. Satellite images also often show that the eye is completely clear of clouds, with a view right down to the surface of the sea.

The formation of tropical cyclones is governed by a very specific combination of factors. They originate just outside the equatorial trough (about 5–10° farther towards the poles), where the Coriolis acceleration promotes the overall rotation. The sea-surface temperature must be in excess of 27°C, and there must be little vertical wind shear throughout the depth of the troposphere (which would act to prevent the essentially closed circulation from forming); but there must be divergence at upper levels, helping to draw the humid air up from the surface. Tropical cyclones are low-pressure regions with a warm core, unlike all other low-pressure systems, which have a cold core. The extreme heating arises from the release of latent heat in the rising cloud towers within the spiral bands surrounding the centre.

Various features are associated with tropical cyclones. The intense precipitation in the cloud bands is often responsible for a significant fraction – or even the greater part – of annual rainfall in many tropical countries, which are often dependent on such precipitation

Cross references

Beaufort Scale p.154
Convergence p.27
Inter-Tropical Convergence Zone p.18
Latent heat p.48
Wind shear p.166

for their agriculture. The sustained high winds may cause considerable damage, in addition to which the violent updraughts and downdraughts often create subsidiary tornadic vortices, which may add to the destruction. In general, however, the greatest danger comes not from the wind, but from the storm surge associated with the system. The wind stress on the surface of the sea, assisted by the low atmospheric pressure, raises a mound of water beneath the system. Storm surges pass unnoticed on isolated tropical islands surrounded by deep water, but when a tropical cyclone approaches a larger coastline, the gradually shelving sea floor acts to accentuate the height of the surge, which may reach many metres above normal tidal levels and thus cause extensive flooding of coastal areas. The effects may be amplified by the exact nature of the coastline, especially if the surge is funnelled into an estuary, for example. The highest storm surge so far recorded occurred at Hatia Island in Bangladesh on 12 November 1970, and reached the phenomenal height of 12.2m.

Tropical cyclones occur over all the world's major oceans, with (as mentioned) the possible exception of the South Atlantic, where water temperatures are normally too low and there is too great a vertical wind shear for such systems to form regularly.

The tracks of tropical cyclones may be very erratic, but they generally move westwards and gradually towards the poles at speeds that are typically about 10 knots (19kph). If they reach latitudes of 20–30°N or S they frequently display dramatic recurvature, a sudden change of direction to the north-east or south-east (in the northern and southern hemispheres respectively). The systems generally decay when their primary source of energy decreases, that is, when they move over land or over cooler waters. The remnants of the tropical cyclone may then continue to higher latitudes and become depressions (extratropical cyclones) or else merge with existing low-pressure systems. In the latter case, they may produce a dramatic intensification of the pre-existing system.

As with any weather system, tropical cyclones do not emerge fully grown, but go through a number of stages of growth. Atlantic hurricanes often begin as an easterly wave,

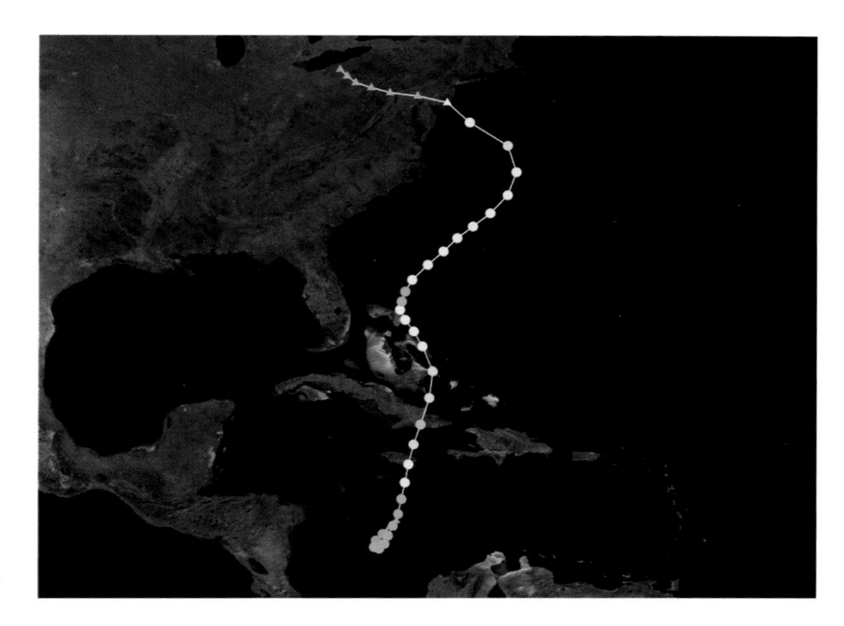

ABOVE The track of the system that came to be known as Superstorm Sandy, which was regarded as a hurricane over only part of its track. In this particular case, the system did not show the recurvature characteristic of most hurricanes, and which would have carried it out over the central Atlantic rather than devastating the eastern seaboard of the United States. *(National Weather Service)*

BELOW The tropical disturbance that later became Hurricane Sandy, photographed by NASA's Terra satellite on 20 October 2012. At this stage, the system was just a cluster of individual thunderstorms. *(NASA)*

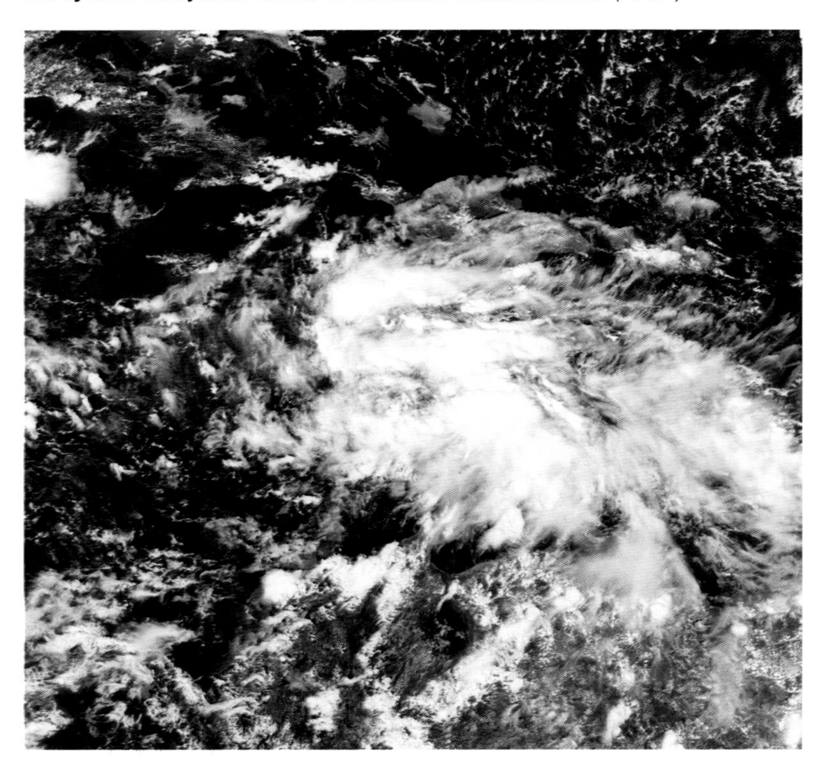

ABOVE A slightly unusual grouping of three typhoons, close together over the western Pacific on 7 August 2006. The youngest, Bopha (centre left) strengthened to become a tropical storm just hours before the image was obtained. Typhoon Maria (top right) is a day older, and shows distinct spiral structure and the suggestion of a central eye. Both Bopha and Maria were similar in size and strength, with sustained winds of approximately 90 kph and 100 kph, respectively. The oldest system, Typhoon Saomai (bottom right), yet another day older, had sustained winds of about 140 kph. It is obviously a far larger and more developed system than Maria, with distinct, long spiral bands and a clearly visible central eye. *(NASA)*

SAFFIR-SIMPSON SCALE

The assessment is in terms of both the wind speed and possible storm-surge height. Note that the scale is defined in non-metric units.

Category	Central pressure		Wind speed		Storm surge	
	in	hPa	mph	kph	ft	m
1 Weak	> 28.94	> 980	74–95	119–153	4–5	1.2–1.5
2 Moderate	28.50–28.91	965–979	96–110	154–177	6–8	1.8–2.5
3 Strong	27.91–28.47	945–964	111–130	178–208	9–12	2.8–3.7
4 Very strong	27.17–27.88	920–944	131–155	209–251	13–18	4.0–5.5
5 Devastating	< 27.17	< 920	> 155	> 252	> 18	> 5.5

also known as a tropical wave – a large-scale shallow trough, usually more marked at upper levels than at the surface, which travels westwards in the zone covered by the north-easterly trades, and which is accompanied by a marked increase in cloud cover and precipitation. The easterly wave then develops into what is known as a tropical disturbance, an area of organised convection that is non-frontal and associated with weak low pressure, light winds, increased cloud cover and light precipitation. This may, in turn, develop into a tropical depression, which is again non-frontal, but consists of a low-pressure area with closed isobars and circulation. Such systems develop, in particular, where there is significant convergence, especially at the Inter-Tropical Convergence Zone. Wind speeds are relatively low, being less than 18m/s (approximately Force 7 on the Beaufort Scale). Although most systems do not develop any further, a few may move on to the next stage, which is that of a tropical storm.

A tropical storm is a well-organised circulation around a low-pressure centre, with maximum sustained (one-minute) wind speeds of 18–32m/s. The curved cloud bands are now distinctly visible in satellite images. It is at this stage that the system is usually given a distinctive name. Such storms may be subdivided and described as moderate (wind speeds 18–25m/s) or severe (wind speeds 26–32m/s), approximately Force 8–9 and 10–11 respectively on the Beaufort Scale. Such tropical storms may then develop still further and become fully fledged tropical cyclones. (Conversely, of course, tropical cyclones that weaken may still be strong enough to be classified as tropical storms.)

Although there are some differences in the definitions of tropical storms and cyclones, and the naming procedures, depending on the ocean basin and the meteorological offices involved, the classification used for North-Atlantic and Eastern Pacific systems and assigned by the National Hurricane Center of the United States National Weather Service may be taken as an example. This is the Saffir-Simpson Scale, developed by a civil engineer (Saffir) and a meteorologist (Simpson) in 1971.

Chapter 12

Light, colours and optical phenomena

Light and colour in the atmosphere

The colours of the sky, clouds and other objects provide information about the current state of the atmosphere, and some provide indications of how the weather may develop.

The normal blue colour of the sky arises because of the effect of the oxygen and nitrogen molecules that make up the bulk of the atmosphere. These molecules are of such a size that they preferentially scatter the short-wavelength blue light from the Sun in all directions, but have little effect on longer wavelengths. (Violet light is also strongly scattered, but this is largely invisible to human eyes.) This blue scattering is present all the time, as shown by long-exposure photographs taken at night, in which the sky appears precisely the same tint of blue as photographs taken during the daytime.

This scattering of blue light is why the rising or setting Sun (or Moon) usually appears orange or red: all shorter wavelengths have been scattered aside during their path through the atmosphere and therefore do not reach the observer. The longer the path, the greater the amount of scattering. It is for this reason that when the sky is covered by a thick layer of cloud that does not extend to the horizon, the clear sky seen in the distance appears to have either an orange or a reddish tint.

When the Sun is higher in the sky, rather than right on the horizon at sunrise or sunset, some of the sunlight's orange and yellow wavelengths may remain to illuminate clouds and other objects. This is seen in the sequence of different colours seen on mountain peaks at sunset that is known as alpine glow or Alpenglüh (German for 'alpine glow'). The peaks are illuminated in sequence by yellow, pink, red and purple

ABOVE At sunset only the orange and red wavelengths of light remain to illuminate the clouds. *(Author)*

FAR LEFT Alpine glow (Alpenglüh) on nearby mountain peaks, photographed from the Wendelstein mountain in the Bavarian Alps. *(Claudia Hinz)*

LEFT Cumulonimbus in the eastern sky, illuminated by the setting Sun. *(Author)*

LEFT The twilight arch immediately after sunset, showing the yellow tints just above the horizon. *(Author)*

CENTRE With the Sun farther below the horizon, purple tints begin to appear in the twilight arch. *(Author)*

BELOW The purple light, recorded following the eruption of Mount Pinatubo in 1991, using a film (no longer available) that was able to record the distinctive tint. *(Richard Barton)*

light, and may appear very prominent, because lower ground is already in darkness. A similar effect is seen on high clouds in the east (usually Cumulonimbus). Naturally the reverse sequence of colours is also seen at sunrise.

The sky itself goes through a sequence of colours in what is called the twilight arch above the point where the Sun has set. Shortly after the Sun's disc has finally disappeared the area close to the horizon takes on a yellow tint. Above that is an area often described as being salmon pink, and even further above there is a pale blue region that gradually blends into the darker blue high above. Somewhat later, the yellow has disappeared to be replaced by orange, while higher in the sky the pink grades into purple and that into the very dark blue overhead. Naturally, a reversed sequence of similar colours is visible just before sunrise.

The purple light

On rare occasions following major volcanic eruptions, the whole of the western sky becomes a vibrant purple shade and bathes the landscape in an unusual, and slightly eerie light. The appearance is so distinctive that it is difficult to mistake for the normal sunset tints in the sky. The effect arises when major eruptions eject sulphur-dioxide droplets high into the stratosphere. There they combine with water to form sulphuric acid droplets that scatter light of different wavelengths to normal. These mix with the usual blue scattering to produce the utterly distinctive purple light. The effect is quite rare, and the last major display was associated with the gigantic eruption of Mount Pinatubo in the Philippines in 1991. It used to be very difficult to photograph the purple light, because of the way in which colour films recorded the various spectral colours. Only one or two types of film would record it properly. It seems that digital cameras may be more successful, but there has not yet been an eruption comparable

RIGHT **Purple and orange tints in the sunset sky following the eruption of Mount Pinatubo (Philippines) in 1991.** *(Strach)*

CENTRE **Distinctive striations in a layer of volcanic dust, ejected by El Chichón (Mexico) in 1985.** *(Author)*

to that of Mt Pinatubo, where the ejected material in the stratosphere spread right around the globe and also to high latitudes.

Volcanic eruptions do, of course, sometimes eject large quantities of volcanic ash and dust into the upper atmosphere. These, like the purple light, may give the sky a distinctive colour and appearance. Generally the twilight arch is broadened and the yellow and orange shades predominate. When a layer with particularly large amounts of material is present, the orange colouration may extend well across the sky. It tends to show slight variations in tint where there are undulations in the layer or thicker and thinner regions. Such ripples generally appear to run across the sky, roughly at right angles to the light from the Sun.

Earth's shadow

Because it is a solid object, the Earth itself casts a shadow on to the atmosphere and out into space. As the Sun sinks in the west at sunset, the Earth's shadow slowly rises in the east. When conditions are favourable (usually when there is some haze or dampness in the atmosphere) the shadow is clearly visible as a steely-grey (blue-grey) band along the horizon, with its upper border displaying a reddish tint. This red border is known as the counterglow and occasionally as the gegenschein, but the latter term is best reserved for an astronomical effect (caused by interplanetary dust) where a faint patch of light is seen in the heavens at the antisolar point on very clear, dark nights.

As the Sun sinks lower in the west, the Earth's shadow darkens and rises in the east until it eventually disappears, merging with the dark sky above. A similar effect is, of course, seen at sunrise, although at that time of day it tends to be less distinct because the air is usually cleaner, so the shadow is weaker and more difficult to see.

BELOW **The steel-grey Earth's shadow, bordered by the reddish counterglow, rising in the east as the Sun sinks in the west.** *(Author)*

RIGHT The striking shadow of Mt Fuji cast at sunrise onto a mixture of Stratus and Stratocumulus clouds. *(Barbara Becker)*

RIGHT The shadow cast by Mauna Loa in Hawai'i at sunset may be seen to the left of the dome of the Canada-France-Hawai Telescope. *(Steve Edberg)*

Cross references

Atmospheric composition p.14
Brocken Spectre p.133
Haze p.104

Anyone who is near the peak of a mountain at sunset or sunrise has the chance to see the mountain's shadow stretching, like a broad cone of shadow into the distance, either cast on to the atmosphere or on to a lower layer of cloud. Such mountain shadows may be extremely striking, and some, such as that cast by Mount Fuji in Japan, have become notable spectacles, with people making the trek to the summit – especially before sunrise – to see the effect for themselves. Whatever the actual shape of the mountain's peak, a mountain shadow always appears as a dark triangle, because what are actually parallel sides of the shadow seem to converge through the effects of perspective. (Such convergence of shadows is, of course, partly responsible for the Brocken Spectre illusion mentioned later.)

Crepuscular rays

Frequently, rays of light or bands of shadow are visible, apparently radiating from a single point in the sky. These are known as crepuscular rays and were originally named because they were visible in twilight at dawn and dusk. In fact, similar rays are often seen at other times of the day.

These rays are most clearly visible when the air is slightly hazy or full of water vapour. The original form appears when the Sun is just below the horizon at sunset or sunrise, and distant mountain peaks or clouds cast shadows across the sky. The dark shadows are often said to show a slight greenish tinge, which probably arises because of the contrast between the shadows and the bright rays of sunlight that are reddened by their passage through the atmosphere.

Crepuscular rays are often seen when sunlight penetrates through the gaps between the elements of Stratocumulus (in particular). This form has been given many different names over the years, being called 'Jacob's Ladder', 'Apollo's backstays' (by seamen), and 'the Sun drawing water'. Because this form often displays many narrow bands of shadow, it has sometimes been mistaken for distant rain, even though, as we have seen, heavy rain does not occur with Stratocumulus clouds.

BELOW Crepuscular rays cast by very distant Cumulus clouds when the Sun was well below the horizon at sunset. *(Author)*

BELOW 'The Sun drawing water': shafts of light and narrow shadows cast by sunlight filtering through a layer of Stratocumulus. *(Author)*

A slightly different form of crepuscular rays is seen when the Sun, in any part of the sky, is hidden by a cloud, but shafts of light pass through gaps in the cloud's outline, in between the bands of shadow. This type is usually seen when the air is particularly hazy or humid.

Occasionally, and dependent on the conditions being just right, crepuscular rays extend right across the sky and, through the effect of perspective, seem to converge on the antisolar point on the opposite side to the Sun. These are known as 'anticrepuscular rays', and on very rare occasions may form a particularly dramatic display.

LEFT Dramatic shadows cast by a fine display of Altocumulus floccus clouds. *(Author)*

LEFT Crepuscular rays surrounding a Cumulus congestus cloud with 'a silver lining' and a dark shadow halo, photographed on an extremely humid, sultry day in Washington DC. *(Author)*

Optical phenomena

Optical phenomena that are visible in the sky are only indirectly linked to the weather, but they give useful clues to both its current and future states. In addition, some of them are directly related to the type of particles (water droplets or ice crystals) that are present in clouds or even in an apparently clear sky.

Most of the various optical phenomena that are visible in the atmosphere (together with some that occur on the ground) may be divided into two broad categories: those created by water droplets, which include some effects visible on the ground, and those produced by ice crystals. We may also consider one or two other effects, such as refraction (which is involved in mirages), under the same general heading.

Water-droplet effects

- Rainbows
- Fogbows
- Dewbows
- Glories
- Heiligenschein
- Coronae
- Iridescence

Of these, all but the last two, iridescence and coronae, are visible in the direction that lies away from the Sun (or, in some cases, the Moon).

Rainbows

Rainbows are familiar to everyone and are, of course, most commonly seen in showery weather

BELOW A fine display of anticrepuscular rays, photographed near Buford, Wyoming (west of Cheyenne), with the shadows probably being cast by the peaks of the Medicine Bow range. *(Nate Cassell)*

RIGHT A double rainbow, photographed in New South Wales. This display is a little unusual in that the area between the primary and secondary bows (Alexander's Dark Band) is no darker than the sky outside the secondary bow. *(Duncan Waldron)*

when both rain and sunshine are present at the same time. (The Moon does produce similar bows, but its light is much weaker than that from the Sun so colours are rarely visible in such bows.) The form most commonly seen is the primary rainbow, which occurs centred on the antisolar point – the point on the sky directly opposite the Sun, relative to the observer's head (or the camera) – with the red arc on the outside (radius 42°) and the violet edge on the inside (radius 40°). It occurs with a single reflection from the rear of the falling raindrops, with the colours being created by refraction (dispersion) within each individual raindrop. Frequently, of course, only part of the bow is seen, either because rain is not falling in part of the sky, or because sunlight is prevented from falling on the raindrops by an intervening cloud.

Secondary rainbows are fairly common. Again, they are centred on the antisolar point but with a reversed order of colours (that is, with red on the inside, radius 52°, and violet on the outside, radius 54°). They arise through two reflections within each raindrop. The area between the primary and secondary bows appears significantly darker than the surrounding sky. This area is known as Alexander's dark band, and occurs because within it, light from the Sun is actually reflected away from the observer.

The higher the Sun in the sky, the lower the top of the rainbow. When the Sun's elevation is greater than 42° or 54°, the primary and secondary bows, respectively, become invisible. Conversely, when the Sun is on the horizon, at sunrise or sunset, the bows form perfect semicircles. Under such circumstances, generally only the red colour is seen, all other colours having been scattered away by the atmosphere. Complete, fully coloured, circular rainbows may sometimes be seen from aircraft under suitable conditions.

Occasionally, pastel-coloured bands are visible within the primary bow. These are known as supernumerary bows (or interference bows)

RIGHT A particularly brilliant primary rainbow, photographed from the Wendelstein mountain in the Bavarian Alps. The violet colouration is normally invisible to the naked eye because of the limited range covered by human vision. *(Claudia Hinz)*

RIGHT A low primary rainbow with supernumerary arcs, produced by a high Sun. *(Author)*

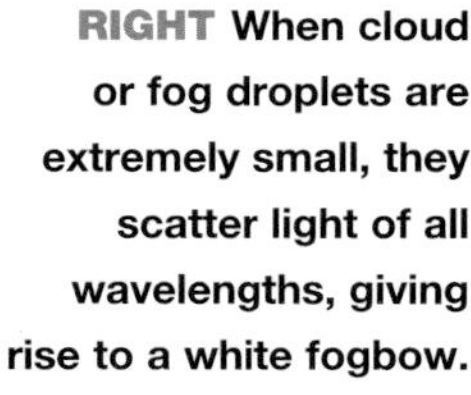

RIGHT When cloud or fog droplets are extremely small, they scatter light of all wavelengths, giving rise to a white fogbow. *(Claudia Hinz)*

and are created by light that has taken slightly different paths through the raindrops. The colours usually appear violet and pale green to the naked eye.

Additional bows may appear if there is a reflecting surface behind the observer (such as a calm lake or even a large expanse of glass in glasshouses). The reflected light may create additional primary or secondary bows, but these will be centred on a point that is higher in the sky.

Large raindrops produce brighter bows, in which the red colouration may be particularly striking. With smaller drops the red is less prominent, and the spacing between any supernumerary bows increases. With very tiny drops the colour disappears completely and the bow becomes a white arc, known as a fogbow, which may be accompanied by a white supernumerary bow.

Dewbows

Coloured bows sometimes occur on dew-covered grass. They arise from exactly the same process as rainbows, but because the droplets of dew are lying on a more or less horizontal surface – rather than being raindrops falling vertically through the air – a dewbow appears as an ellipse or a hyperbola, and generally only a portion (rather than a full arc) is visible. Such bows tend to be most conspicuous in autumn, when dewdrops hang on spiders' webs that are stretched horizontally between blades of grass.

Glories

A glory is a series of coloured rings that surrounds the antisolar point. Such a glory is most commonly seen against mist, fog or a cloud bank, surrounding the shadow of the individual observer's head. An observer only sees the rings surrounding the shadow of their own head, not around those of any companions. The rings are very similar to those of a corona, with violet on the inside and red on the outside. Multiple rings may occur and the radii of the coloured rings are inversely proportional to the size of the water droplets: the smaller the droplets, the larger the radii of the rings. Nowadays glories are very commonly seen from aircraft. If the plane is flying close to the cloud, its shadow is often visible in the centre of the glory.

The Brocken Spectre, sometimes seen at the same time as a glory, is an apparently enlarged shadow of the observer cast on a bank of mist, fog or cloud. It is actually an optical illusion,

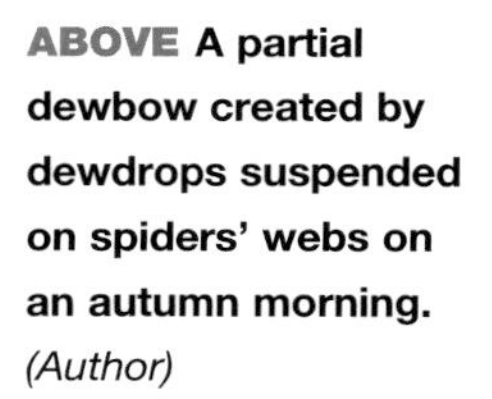

ABOVE A partial dewbow created by dewdrops suspended on spiders' webs on an autumn morning. *(Author)*

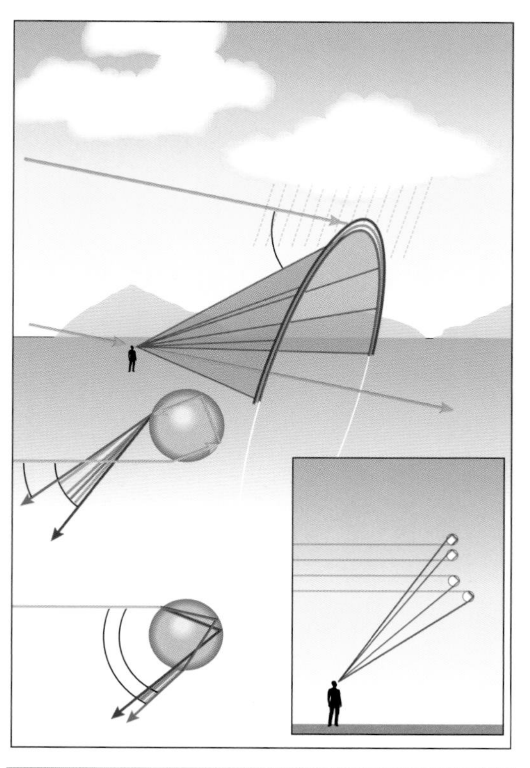

LEFT Rainbows are always seen on the opposite side of the sky to the Sun. A single reflection within the raindrops (a) produces the inner (primary) bow, and two reflections (b) the outer (secondary) bow, with a reversed order of colours, and in a different position in the sky (c). *(Ian Moores)*

LEFT A brilliant glory on mountain mist around the shadow of the observer's head. *(Claudia Hinz)*

RIGHT The white heiligenschein surrounding the shadow of the observer's head (and of the camera) on dew-covered grass. *(Steve Edberg)*

related to the tunnel illusion, and is particularly strong when nearby objects (such as rocks) cast lines of shadow that appear to converge through perspective. The observer's shadow is apparently more distant (and thus seems larger) than otherwise expected.

Heiligenschein

The heiligenschein (German 'holy light') is a bright colourless halo of light around the shadow of the observer's head on the ground (*ie* around the antisolar point). The strongest effect occurs when sunlight falls on dew-covered grass. The drops of water and the blades of grass act as retro-reflectors, returning the sunlight back towards the source. The same colourless halo may be observed with bright moonlight.

Cross references

Measuring angles on the sky p.158
Nacreous clouds p.78
Sky colours p.127

RIGHT The inner aureole of a corona in thin, water-droplet Altocumulus clouds. *(Author)*

RIGHT The outer rings of a corona are often strongly coloured and their shape may be slightly distorted by the structure of the cloud in which they appear. *(Author)*

There is a somewhat similar effect, although not created by water droplets, known as the 'hot spot' to aerial photographers. This is most commonly seen from an aircraft, where a bright spot of light appears to glide across the ground at the antisolar point. The same effect may be seen by an observer on the ground, where there is a bright patch of light around the shadow of the observer's head on a rough surface such as grass or leaves. This simply arises because the observer is looking down-Sun and every visible surface is brightly illuminated, and the shadows of the blades of grass or leaves are hidden by the objects themselves. Further to the side, away from the down-Sun direction, the shadows begin to become visible, so the surrounding area appears darker.

Coronae

A corona is an effect where one or more series of coloured rings appears around the Sun or Moon. The effect is more often seen around the Moon, because the stronger sunlight tends to make the phenomenon difficult to see. A corona arises through diffraction by water droplets in thin stratiform cloud. A complete display consists of an inner aureole (a bluish-white disc with a brownish-reddish outer edge) and an outer set (or sets) of coloured rings, with violet on the inside and red on the outside. Uniform drop size produces the purest colours, but when there is a mixture of droplet sizes the inner aureole is often all that is visible. The radius of the rings is inversely proportional to the size of the water droplets. In many cases, because the clouds are broken rather than a complete even layer, only parts of the rings are visible. Even so, the visible portions may still be surprisingly bright, with clear colours.

Iridescence

Iridescence – which is also known as 'irisation' (from Latin 'iris' or 'rainbow') – arises through the same mechanism as coronae (*ie* diffraction of light by water droplets). It appears as often brilliantly coloured bands of light that tend to run parallel to the edges of clouds that are approximately 30–35° from the Sun. Red and green tints are those most commonly seen, although both yellow and blue may sometimes appear. The most pure (and strongest) colours appear when there is very little variation in the size of the cloud droplets. Iridescence, like coronae, is a sign that the clouds involved consist of water droplets rather than ice crystals. As such, both effects are most commonly seen in Cirrocumulus, Cirrostratus, Altocumulus and, occasionally, Altostratus. Iridescence is responsible for the striking colours seen in nacreous clouds.

LEFT Iridescence on the edge of a sheet of Altostratus. *(Author)*

Ice-crystal effects

- 22° halo
- Parhelion
- 46° halo
- Circumzenithal arc
- Parhelic circle
- Sun pillar
- Subsun

These effects are listed in approximately the order of frequency with which they are seen.

22° halo

The most commonly seen ice-crystal effect is the 22° halo. As its name suggests, it is a ring, 22° in radius, surrounding the Sun (or Moon). It often appears in the thin layer of Cirrocumulus that covers the sky ahead of an approaching warm front, when it occurs about once every three days at temperate latitudes. Although it is generally seen as white, on occasions it will show a slight tinge of colour, with red at the inner edge and violet at the outer.

Parhelion

Occurring almost as frequently as the 22° halo are parhelia (singular 'parhelion'), also known as 'mock suns' or 'sun dogs'. These often occur at the same time as the 22° halo, and are bright spots that seemingly lie at the same distance from the Sun. In fact, the exact position of a parhelion depends on the altitude of the Sun, so it sometimes lies somewhat more than 22° from the Sun. In addition a parhelia often show a bright, white 'tail' that points away from the Sun. When bright, parhelia show a complete range of spectral colours and may be extremely striking, especially when they appear in a patch of cirriform cloud and unaccompanied by the 22° halo. A similar phenomenon may appear with moonlight, when such a spot of light is known as a 'paraselene' (or 'mock Moon'). Because moonlight is so much weaker than sunlight, colours are rarely seen in paraselenae, but the phenomena occur in the same types of cloud as parhelia.

46° halo

There is another halo ring that occurs surrounding the Sun, this time with a radius of 46°. Again, it may show faint colours, red on the inside and violet on the outside. It is normally visible only in Cirrostratus, its light being weaker than that of the 22° halo, so is rarely detectable in isolated patches of cirriform cloud.

LEFT A fine halo display in Cirrostratus cloud, showing the 22° halo, a portion of the 46° halo, parhelia on both sides of the Sun, an upper arc of contact with the 22° halo, and a circumzenithal arc. *(Dave Gavine)*

LEFT A bright, coloured parhelion in a patch of Cirrus, with an indication of the bright 'tail' that is sometimes seen, pointing away from the Sun. *(Author)*

ABOVE A circumzenithal arc, seen in Cirrus uncinus and a thin veil of Cirrostratus. *(Author)*

Circumzenithal arc

One of the brightest halo effects is the circumzenithal arc. This consists of an arc, usually about 120° in length, that is centred on the zenith – the point directly above the observer's head – and is usually symmetrical about the line joining the zenith and the Sun. It shows brilliant spectral colours and is, unfortunately, often described in the media as an 'upside-down rainbow'. This is, of course, completely wrong, because it is caused by light passing through ice crystals, not raindrops.

Parhelic circle

With extensive Cirrostratus cloud (in particular) another effect may appear. This is the parhelic circle, a bright arc that extends around the sky at exactly the same altitude as the Sun, and thus parallel to the horizon. A complete 360° arc is uncommon, depending as it does on there being ice crystals all round the observer, but shorter arcs are moderately frequent.

Sun pillar

A moderately common phenomenon is a sun pillar. This arises when the ice crystals are in the form of flat plates that are floating more or less horizontally in the air. Light from the Sun is reflected by the flat surfaces, producing a vertical streak of light above (and below) the position of the Sun.

Subsun

When the observer is in a relatively high location (such as on a mountain or in an aircraft) it is sometimes possible to see a subsun (also known as an 'undersun'). This is an elliptical patch (with its long axis vertical) that appears at the same distance below the horizon as the true Sun is above it. Like the sun pillar, it is caused by the reflection of light from the more or less horizontal faces of flat hexagonal plates

RIGHT A sun pillar in an extensive sheet of striated Cirrus above the setting Sun. *(Author)*

of ice. On rare occasions, coloured subsuns (subparhelia, subsun dogs) may appear at approximately 22° on each side of a subsun, much as parhelia appear relative to the actual Sun.

Other halo effects

There are many additional halo effects that may occur when ice crystals are in the sky – far too many to describe in detail here. These effects may take the form of partial coloured arcs and 'chevrons', some of which touch the top of the 22° or 46° haloes; and white arcs that contact different portions of the normal circular haloes, or intersect the parhelic circle, producing bright spots of light where they cross it. One of the most striking effects is the circumhorizontal arc. This is the brightest and most colourful halo phenomenon, which sometimes exhibits spectacular colours. It lies parallel to the horizon but appears only when the altitude of the Sun is greater than 58°, so is most readily seen from low latitudes. It lies just below the 46° halo, if this is visible, and becomes tangent to it when the Sun's altitude is 68°. Other effects include:

- Haloes with unusual radii: circular arcs with radii of 9°, 18°, 20° and 35° have been reported.
- Parry arcs: bright arcs above and below the 22° halo; very variable in shape.
- Lowitz arcs: arcs from the 22° halo, below the parhelic circle, extending upwards to the altitude of parhelia; very variable with solar elevation.
- Supralateral and infralateral arcs: arcs to each side and above and below the 46° halo; also very variable with solar elevation.

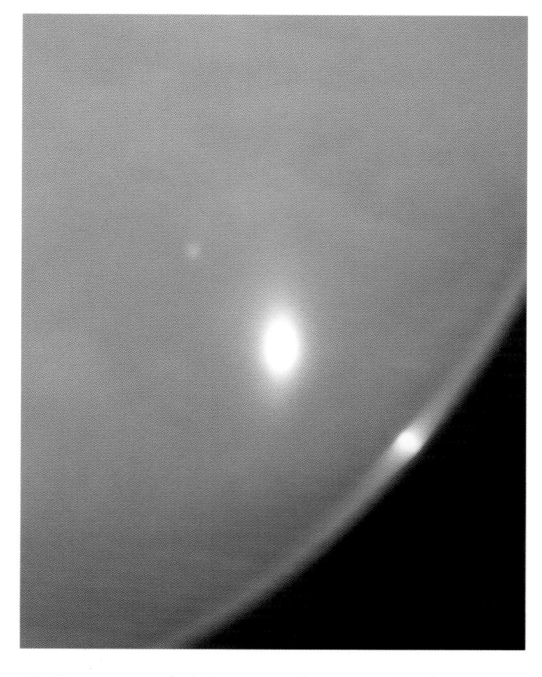

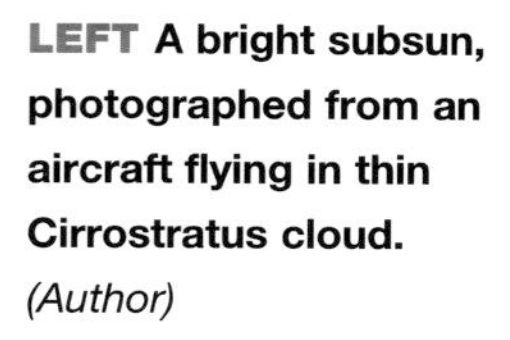

LEFT A bright subsun, photographed from an aircraft flying in thin Cirrostratus cloud. *(Author)*

Cross references

Diamond dust p.97
Ice-crystal forms p.135
Measuring angles on the sky p.158

LEFT A brilliant subsun and partial 22° halo, photographed from a high vantage point so the ice crystals are below the observer. *(Claudia Hinz)*

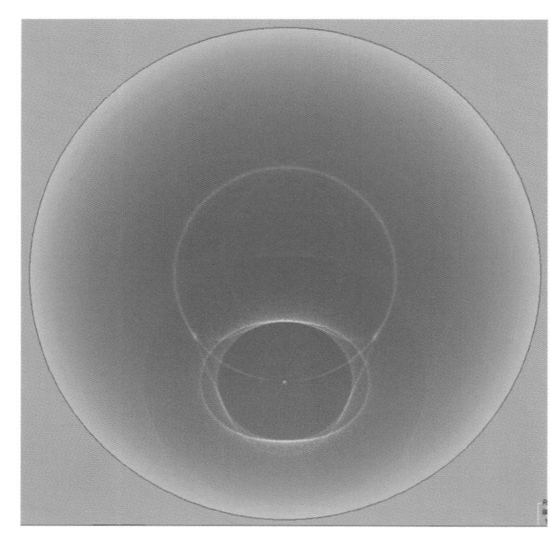

RIGHT A simulation, using the HaloSim program, of a famous complex halo display over St Petersburg in 1790, seen and described by Tobias Lowitz. *(Author)*

RIGHT The setting Sun may assume many different shapes. This form is often known as 'the Vase'. The base on the horizon often expands into the 'Omega' shape. *(Maria Zinkova)*

BELOW The green flash, with the Sun setting over the Pacific. The magnified images show that yellow and blue tints also occur, although the areas are too small to be distinguished by the naked eye. *(Wikimedia)*

- Anthelion: a bright spot at the antisolar point (*ie* directly opposite the Sun).
- 20° parhelia: bright spots on the parhelic circle, 120° from the Sun.
- Anthelic pillar: vertical pillar centred on the anthelic (or antisolar) point.

Many halo effects are particularly strong when the air is full of the minute ice crystals known as 'diamond dust'. Such conditions are particularly common in the Arctic and Antarctic, and scientists working at the South Pole have discovered many unusual and rare forms of arc.

Because of the fixed angles between the faces of ice crystals it is possible to run computer simulations of numerous aspects of halo displays. The image shows one such simulation of a famous display over St Petersburg in 1790.

Refraction and mirages

Refraction always exists in the atmosphere, and rays of light are deviated to differing degrees depending on the density of the air through which they pass. Under normal circumstances the density increases towards the surface, but any effect is not readily apparent. When the Sun's disc appears to touch a sea horizon at sunset or sunrise, the whole of the disc is actually below the horizon, and refraction in the atmosphere has 'lifted' it into view above the geometrical horizon. The disc often appears slightly flattened, and this is because the light from the lower limb (edge) of the disc passes through denser air than light from the upper limb and is deviated to a greater degree. The same effects occur with the Moon. Frequently the disc is distorted into strange shapes, because the light from various parts of the disc passes through layers of different densities and is refracted to various degrees. Mirage effects (described shortly) may also produce strange distortions of the disc, including the form known as an 'omega sun', where an inverted image produces a vase-like shape or one like a Greek letter omega.

The green flash is also a refraction effect, where the last portion of the setting Sun appears a bright emerald green. A similar effect also occurs at sunrise. Very occasionally, a larger green segment is seen and, on extremely rare occasions, the uppermost portion of the Sun appears bright blue.

Mirages occur when there are unusual density gradients in the atmosphere, produced by variations in temperature or humidity. The changing density causes differential refraction, which affects the position or appearance of distant objects. These may be distorted or, in true mirages, may create inverted or additional images.

There are two basic forms of true mirage: inferior mirages where an object appears 'lower' than it would when seen through a homogeneous layer of air, and superior mirages where the object appears higher. The most commonly noticed mirage is an inferior mirage where pools of water appear to be lying on a heated road surface. The 'pools of water' are actually an image of the sky, light from which has been sharply curved by the hot, low-density air above the road. The same effect occurs over any heated surface, such as a desert, and under certain conditions an inverted image of distant objects (vehicles or trees, for example) may be visible.

In a superior mirage, there is a strong temperature inversion with a layer of warm air above a much colder one. Such conditions often occur over the sea during a hot day, or in spring, when the water retains its low wintertime temperature. They also occur over a large expanse of ice. Distant objects appear to be floating in the air, often as an inverted image, and multiple images are common. In one particular effect, known as Fata Morgana, distant objects are greatly elongated and give the appearance of a cliff, tall buildings or whole cities in the distance. This is usually a highly distorted image of the surface of the sea or ice floes.

Related to mirages are other effects where the visibility of distant objects is affected or where objects appear elongated or compressed. Any (or several) of these effects may be present at the same time as inferior or superior mirages, if there are multiple layers of air with different densities and rates at which the densities change. This may lead to highly complex images of distant objects.

- Looming: object raised above its customary position and thus seen at a greater distance than normal (air density decreases unusually rapidly with height).
- Sinking: object depressed below its customary position and thus part or all of the object, normally seen, is invisible (air density decreases more slowly with height than normal, or even increases).
- Stooping: object compressed vertically (layer closest to the surface has a higher temperature and lower density than normal).
- Towering: object elongated vertically (more rapid decrease in density with height than normal).

ABOVE An inferior mirage, with the inverted image of a water tank appearing to be reflected in a pool of water. Photographed over the hot expanse of Edwards Dry Lake at Edwards Air Force Base in California. *(Steve Edberg)*

Cross references

Atmospheric layers p.10
Composition of the atmosphere p.14

BELOW Superior mirage of Point Reyes, seen from San Francisco, when a temperature inversion was present. *(Wikipedia Commons)*

PART THREE

Observing the weather and forecasting

OPPOSITE Billows in high Cirrocumulus, partially hidden by the lower Altostratus and even lower Stratus clouds associated with a weak frontal system. *(Author)*

Chapter 13

Observations

Knowledge of the many different processes responsible for creating weather conditions has only come after many decades of observation (and in a few cases from actual experiments). Observations are, of course, key to weather forecasting, which basically consists of three stages: observation of actual conditions; analysis of the observations to determine the state of the atmosphere at any time; and then forecasting.

Meteorological observations are a worldwide enterprise. As we have mentioned earlier, to allow a western-European meteorologist to prepare an adequate forecast for 24 hours ahead requires a knowledge of conditions right the way across the Atlantic, and for one just three days ahead details are needed of the state of the atmosphere over the whole Earth, including the southern hemisphere.

The routine measurements made, worldwide, by the various national agencies include a standard set of observations, established by international agreement through a specialised agency of the United Nations, the World Meteorological Organization (WMO), based in Geneva. Meteorological data, obtained worldwide, is made freely available to all participating nations. The overall system is known as the World Weather Watch (WWW), and consists of three primary subsidiary elements:

- **The Global Observing System** – the means by which observations from around the world are made in a standardised form, even though they may originate from many different observational sites, including manned or automatic stations on land; ships; drift or anchored oceanic buoys; aircraft; radiosondes; geostationary and polar-orbiting satellites.
- **The Global Data-Processing System** – the standardised procedures for the receipt, processing, storage and retrieval of meteorological observations.
- **The Global Telecommunications System** – the physical communications network for the rapid collection, transfer and distribution of meteorological data to all parts of the world.

Although the widespread introduction of automatic weather stations (AWS) means that there is considerable freedom in selecting the frequency of observations, the WMO guidelines stipulate that certain key observations should be made simultaneously, on the hour, in accordance with Universal Time Coordinated (UTC). (This time scheme is calculated by the intercomparison of highly accurate atomic clocks maintained by various national standards organisations.) UTC corresponds to the time on the Greenwich meridian, and is not altered by the application of Summer Time or Daylight Saving Time. In meteorological usage, the standard designation for the time zone that is centred on the Greenwich meridian is 'Zulu', abbreviated 'Z', and observations are often quoted as being obtained at '00:00Z', for example.

Surface observations

The surface observations that are normally reported hourly by a manned station include:

- Dry bulb temperature.
- Dew point temperature.
- Mean pressure (corrected to sea level).
- Pressure tendency (change over the last three hours).
- Total cloud amount.
- Cloud type and height of base.
- Current weather.
- Past weather.
- Wind direction.
- Wind speed.
- Maximum gust speed.
- Horizontal visibility.

The methods by which some of these observations are obtained will be described later, as well as the way in which the data are plotted on typical synoptic charts.

Many stations report other observations, especially factors such as precipitation (which is usually given as 12- or 24-hour totals) and sunshine amount (often as daily, weekly or monthly totals). Climatological stations also generally report monthly means for specific elements. Manned stations may report many other observations, such as depth of snowfall, soil temperatures, evaporation rates etc. Certain observations cannot, at present, be made at automatic stations, so these omit details of cloud types and, frequently, horizontal visibility. Reasonably accurate estimates of current and past weather may be made automatically from the appropriate instrumental readings.

The frequency of observation largely depends on the type of observing station. Most civil airports and military airfields make observations every hour, on the hour. Other stations may make observations every 3, 6 or 12 hours. It is, however, essential that regardless of the location of the station, observations are made at 00:00 and 12:00 UTC. Observations made at these times are disseminated worldwide and are used as the primary input data for various numerical forecasting programs. Approximately 10,000 observations are distributed every hour by the Global Telecommunications System to meteorological offices around the world.

The rapid advances in developing electronic means of obtaining meteorological data have meant that observations are now returned by automatic weather stations from inhospitable or relatively inaccessible sites (such as mountain tops, locations in Antarctica and the oceanic buoys moored in the Pacific Ocean to monitor conditions that lead to El Niño events). Some particularly remote stations return observations through orbiting satellites that act as relays. These include hundreds of free-floating buoys that are programmed to sink to a specific depth, return to the surface at intervals to broadcast the observational data, such as temperature, salinity and their location, and then sink automatically back to their pre-programmed depth.

SYNOPTIC CHARTS

The term 'synoptic' is extensively used in meteorology and implies that the data were obtained simultaneously, and thus represent the conditions that applied over the given area at a specific time. Typical synoptic charts are those that show station plots for one of the standard observational times, or analysis charts showing the pressure distribution and frontal systems. The isobaric charts that show thickness levels, mentioned earlier (p.32–37) are also a form of synoptic charts.

ABOVE An automatic weather station being installed near the British Antarctic base of Rothera. *(BAS)*

LEFT A typical deep-water moored meteorological buoy. *(Author)*

ABOVE A typical weather-radar station, in this case the Berrimah station near Darwin in northern Australia. *(Wikipedia)*

Tropospheric observations

Weather radar

Radar systems are extensively used to detect rainfall and to evaluate its amount and direction of motion. The last factor is normally determined by what are termed Doppler-pulse radars, which are able to determine the movement towards or away from the radar from the Doppler shift in the radar echo returned by the water droplets. A typical weather radar will display the pattern of precipitation within a radius of approximately 150km. Such radars are an extremely valuable means of plotting rainfall because of their extensive coverage when compared with the 'spot' readings obtained by rainfall gauges, which have an essentially random distribution across the surface. Radar is, however, limited to determining the precipitation on its way down, rather than the amount that actually reaches the surface, and this is partly, of course, because interference ('ground clutter') from objects on the surface, such as hills or buildings, affects the radar beams at their lowest elevation. Precipitation radars are, however, essential in providing information for nowcasting, because the nature of the images obtained is such that areas of heavy precipitation are immediately obvious, whereas such areas might be completely missed by the network of rain gauges. The data obtained may also be used to determine the total precipitation over a specific period (typically 24 hours). By way of example, the United Kingdom is covered by a network of 12 rainfall radar systems.

RIGHT A wind profiler (in this case, a Swedish installation) using sound pulses (SODAR) to determine wind properties at various altitudes. *(Wikipedia)*

Sferics

The radio waves produced by lightning strokes ('atmospherics' or 'sferics') may be readily detected at great distances, either by conventional direction-finding techniques or by determination of the differences in arrival times at a network of receiving stations. Thunderstorms may be detected at a range of about 5,000km during daytime, and because of different propagation properties at night and for signals travelling towards the east, there is an even greater range of up to 10,000km for storms occurring at night and to the west. The location of the individual lightning strokes may then be used to track the motion of the parent thunderstorm, and such information (particularly from local storms) is regularly used for nowcasting purposes.

Wind profilers

Within the troposphere, measurements of the wind speed and direction at different altitudes may be obtained by wind profilers, equipment that employs radar or sound waves to probe the atmosphere vertically above them. The data obtained are measurements at 1km intervals up to the tropopause (which may be between approximately 8 and 17km above sea level, depending on latitude). At higher altitudes there is insufficient water vapour present to provide an adequately strong return signal. The Doppler

ABOVE An artist's impression of a microburst occurring over an airport, causing extreme wind shear and a major hazard to aircraft, especially on landing. *(NASA)*

shift in the return signal is processed to obtain the wind speed and direction at that specific level. The NEXRAD system that covers the United States is able to track severe storms and is essential in the prediction of the likely formation of tornadoes, microbursts and similar destructive events. More sophisticated ('pencil-beam') Doppler radar equipment (Terminal Doppler Weather Radar – TDWR) systems have been installed at many airports in the United States to detect the intense downdraughts (known as downbursts or microbursts) and consequent wind shear that may be produced by severe thunderstorms and supercells, and which are a particular hazard to aircraft when landing or taking off.

Upper-air observations

Although some observations are reported by commercial flights, the most important data are returned by balloon-borne instrument packages, known as radiosondes, which typically ascend to altitudes of about 21km (roughly 70,000ft), more than twice the general height of airline flight levels. Radiosondes typically measure pressure, temperature and humidity throughout their ascents, giving vertical profiles through the atmosphere, and thus immediately providing the environmental lapse rate, for example. Most modern ones carry GPS receivers, so the location of the sonde is known accurately in all three dimensions

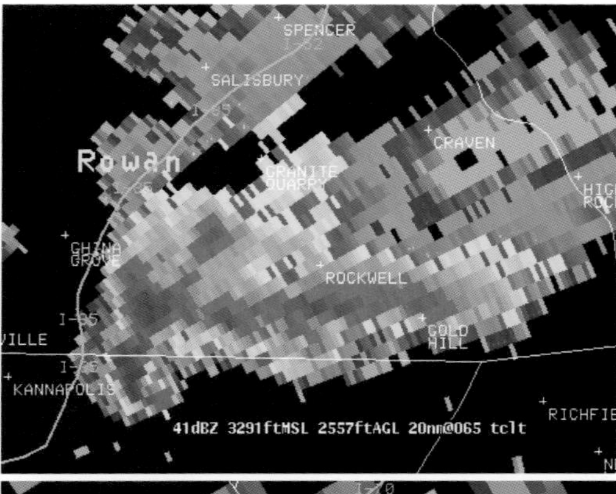

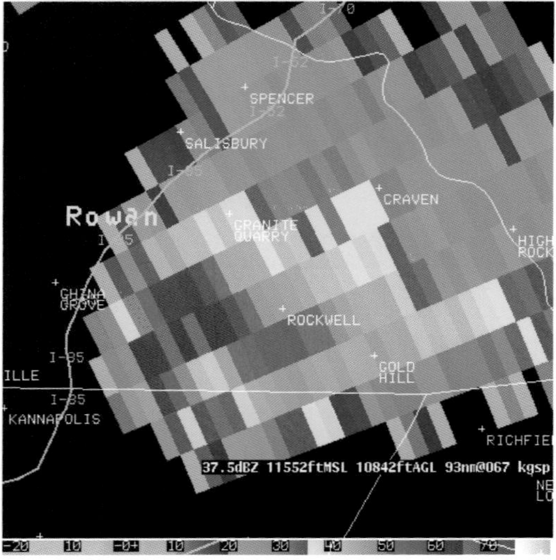

LEFT A comparison of the radar images returned by TDWR (top) and NEXRAD (bottom). Although TDWR has finer resolution, the black areas show where its shorter-wavelength radar has been blocked by intervening heavy precipitation. *(Wikipedia)*

BELOW A radiosonde about to be launched from the old airport at Hilo on the Big Island of Hawai'i. *(NOAA)*

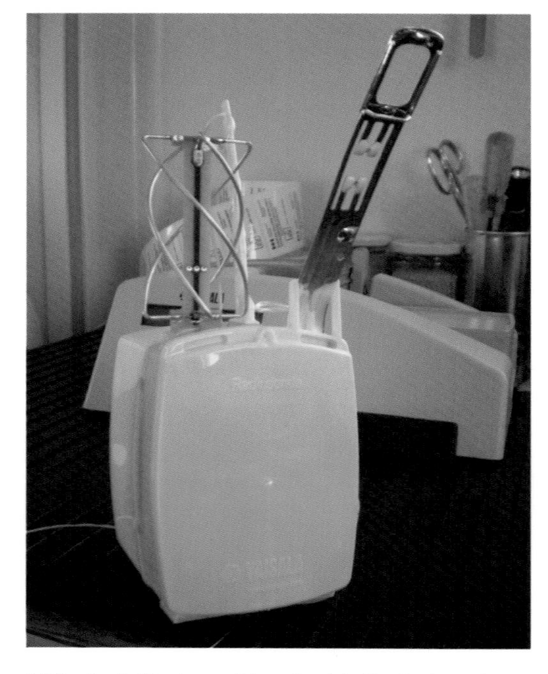

RIGHT **A modern GPS-equipped radiosonde instrument package, ready for launch.** *(Wikipedia)*

(altitude, latitude and longitude). Both the wind speed and its direction may thus be derived easily. Some sondes (known as rawindsondes) are used solely to obtain the wind speed and direction. Other versions may determine the presence of cosmic rays or the concentration of ozone at various levels (ozonesondes).

RIGHT **An artist's impression of the GEOS-8 satellite in orbit. All the GEOS series are of a similar design. Image scanning is performed optically within the body of the satellite.** *(NOAA)*

RIGHT **An artist's impression of a Meteosat Second Generation (MSG) geostationary satellite in orbit. The satellite spins to carry out scanning in the east–west direction.** *(Eumetsat)*

Worldwide there are more than 800 radiosonde launch sites. Most of these are on land, but a number of automated launch facilities are enclosed in modified shipping containers and carried by seagoing vessels. Ascents are always made at the WMO-recommended observational times of 00:00 and 12:00 UTC and, at certain stations, at the additional times of 06:00 and 18:00 UTC. Generally radiosondes are released 45 minutes before the nominal observation time so that vertical profiles are available at the all-important times.

Weather satellites

The data from weather satellites has become indispensable for the preparation of weather forecasts. There are two distinct types of meteorological satellites: polar-orbiting and geostationary.

Geostationary satellites

Geostationary weather satellites orbit the Earth at an altitude of 35,900km above the equator. At this altitude their orbital period is exactly equal to 24 hours, so they remain essentially stationary above a point on the equator. (Because of irregularities in the Earth's gravitational field, the satellites tend to drift out of position, so use thrusters to maintain correct station-keeping. Their lifetime generally ends when all the fuel has been exhausted.) At their location, the curvature of the Earth means that each geostationary satellite has a field of view of about 120°, rather than a full 180°. There is, however, a series of satellites in geostationary orbit that provides essentially complete coverage around the equator. In addition this coverage is more or less continuous. The latest Meteosat Second Generation satellites, for example, return a full-disc image every 15 minutes. The scanning instrumentation on the early satellites provided coverage of just three channels: visible, infrared and water vapour. The latest instruments monitor 12 spectral channels at a much higher resolution.

The images returned in any individual channel are effectively monochrome. The numerical data from one or more channels may, however, be manipulated to produce false-colour images that accentuate the differences

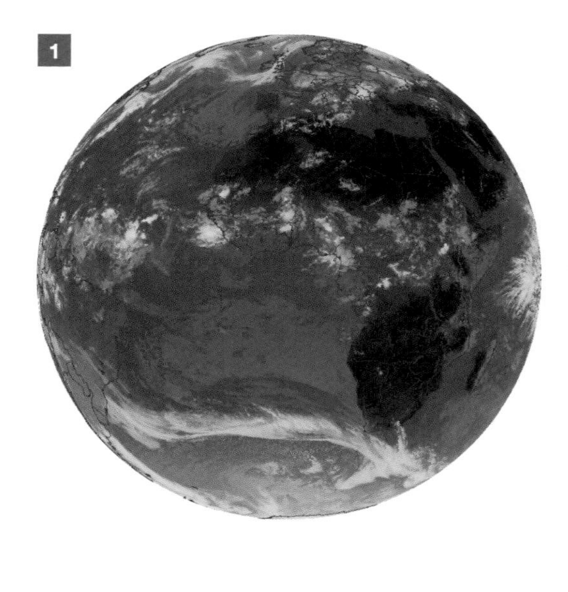

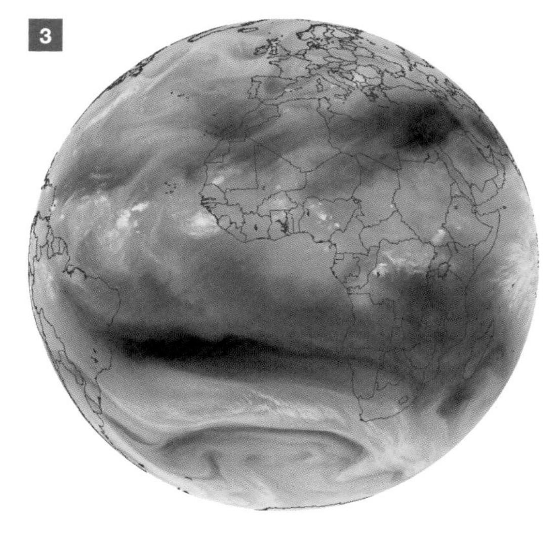

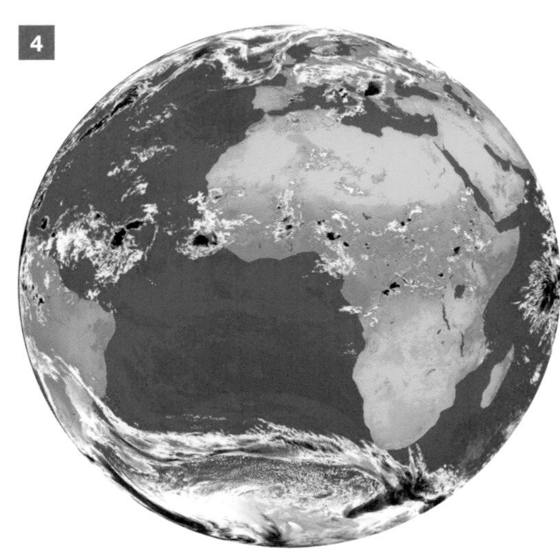

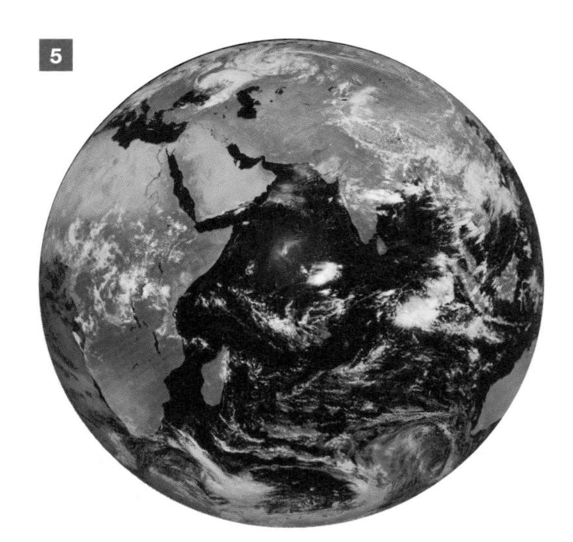

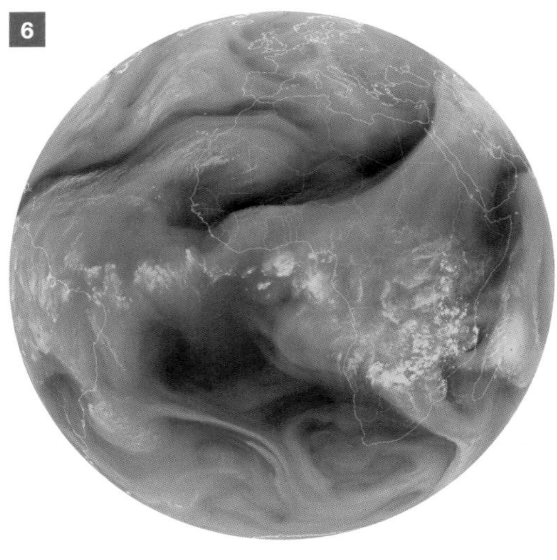

1 An infrared, single-channel image obtained by Meteosat-9 one hour before the visible-light and water-vapour images for the same date. The darkest areas (such as the Sahara and Arabian Deserts) are hottest and the lightest (the cloud tops), are the coldest. *(Eumetsat)*

2 A Meteosat, full-disc, single-channel image in a channel chosen to most closely match the response of the human eye. *(Eumetsat)*

3 A Meteosat image in the channel that best represented the distribution of water vapour in the atmosphere. *(Eumetsat)*

4 A false colour infrared image from Meteosat-9, processed to show the hottest areas as yellow and the very coldest areas as black. *(Eumetsat)*

5 A Meteosat image, processed using multiple channels to give a (false-colour) representation which is approximately the same as that perceived by the human eye. *(Eumetsat)*

6 An older (6 March 2004) image, processed to accentuate the distribution of water vapour in the atmosphere, where the dark areas are the driest. *(Eumetsat)*

in cloud-top or ground temperatures, for example. The 'natural colour' full–disc images are processed in a similar manner to give a close approximation to the colours that would be seen by a human eye.

Polar-orbiting satellites

Because of the curvature of the Earth, the geostationary satellites over the equator are unable to provide detailed coverage of high northern or southern latitudes. The high resolution provided by modern instrumentation means that sufficiently detailed image segments may be obtained up to about 55°N and S, but the polar regions remain inaccessible.

Additional coverage is provided by polar-orbiting satellites. These are launched into highly inclined orbits that take them over the polar regions, and also lie at a very much lower altitude – typically 800–1,000km – than the geostationary satellites. They operate continuously, thus covering the Earth in an unending swathe. The orbits are arranged so that the area of the surface covered by each individual pass overlaps with the preceding pass. The orbits are Sun-synchronous, which means that the satellites pass over a location on the surface at the same time each day, thus providing similar lighting conditions on each pass. In addition, each satellite passes over a particular location twice a day, once on a north–south pass and once on a south–north pass.

Because of their low orbital altitude (and thus superior resolution at the surface) and their continuous coverage of the whole Earth – including oceanic areas where surface observations are sparse – polar-orbiting satellites are ideally suited to monitoring meteorological conditions. Indeed, the very first meteorological satellites were polar-orbiting, although the early images were extremely crude by modern standards.

The longest-running series of polar-orbiting meteorological satellites is the Nimbus series, generally known as the NOAA satellites and operated by NOAA (the United States' National Oceanic and Atmospheric Administration). Other satellites are operated by Russia, China and the European Eumetsat organisation. Many of these satellites broadcast continuously as they cover a swathe of the surface below them. These analogue images, known as APT (Automatic Picture Transmission) are transmitted on frequencies around 137MHz that may be readily captured by amateurs using simple receivers and aerials.

However, the very latest polar-orbiting satellites – the NOAA Polar Operational Environmental Satellites (POES) and the more recent European MetOp satellites – employ much higher frequencies, broadcast in digital form and require more sophisticated receiving equipment and processing software. In addition, computer-controlled receiving aerials are required to track each individual satellite as it passes overhead, unlike the simple, static aerials used for APT reception. Such aerials are more frequently used by amateurs in the United States, because European amateurs are able to take out a one-off licence from the Eumetsat organisation to access the Eumetcast service. A fixed aerial (aimed at a geostationary satellite) then enables them to capture not only processed Meteosat images, but

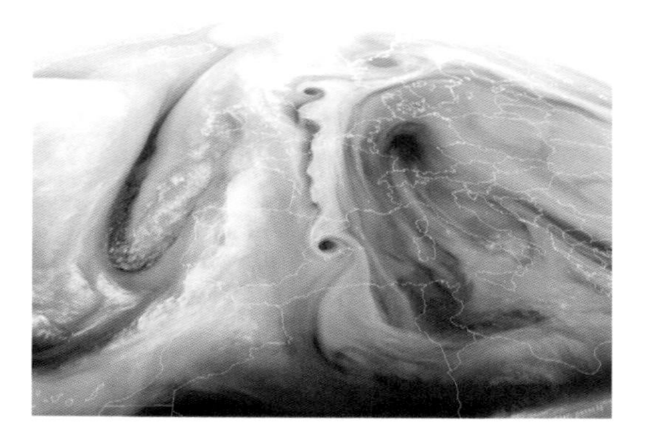

RIGHT A segment from a Meteosat image (channel 5), showing remarkable detail in the distribution of water vapour along a cold front stretching from Finland to the Balearic Islands in the Mediterranean and beyond on 15 January 2006. *(Eumetsat)*

RIGHT The very first meteorological satellite image, transmitted by TIROS-1 (Television Infrared Observation Satellite) on 1 April 1960, the day it was launched. *(NOAA)*

RIGHT An artist's impression of one of the European MetOp satellites in orbit. *(Eumetsat)*

images from other geostationary satellites and the NOAA and MetOp polar-orbiting satellites that are rebroadcast through the satellite.

The MetOp and the NOAA satellites carry a set of instruments in common to provide data continuity. The MetOp satellites use sophisticated instruments to measure atmospheric temperature and humidity to a high degree of accuracy, as well as to obtain profiles of various trace gases (including ozone). Oceanic wind speeds and directions are also determined. As with many modern satellites, both series carry additional equipment to assist with the detection of emergency beacons on ships and aircraft (to trigger search and rescue missions), as well as acting as data relay systems, transmitting data from oceanic buoys and remote automatic weather stations.

Instrumentation

Although many surface meteorological observations are now obtained by electronic means, there are still a large number of observing stations that are equipped with traditional instruments that are read at regular intervals by human observers. Any observing station, manned or automatic, needs to be carefully laid out to ensure that the instruments have the correct exposure and thus return consistent results. Although it is relatively simple to ensure that thermometers (for example) are exposed to a free flow of air, but are at the same time protected from solar radiation by the use of a louvred enclosure (known as a Stevenson screen), the siting of other instruments may require considerable care. Ideally, anemometers and wind vanes (for measuring wind speed and direction) should be carried on a mast 10m high so that they are above the surface boundary layer and its associated turbulence. Even then they need to be sited well away from any buildings or trees that may create turbulence or interfere with the free flow of air. (At many officially maintained automatic stations and on oceanic buoys, where it is impractical for the instruments

LEFT A small automatic weather station at a historic site, Kew Gardens. The data from the various instruments on the solar-powered mast is collected by the data-logger (in the white box) and may be compared with certain measurements by equipment within the Stevenson screen in the background. Note the two forms of rain gauge that are in use. *(David Hawgood)*

LEFT A UK Meteorological Office station (Port Ellen on Islay). Note the mast carrying the anemometer and wind vane at a height of 10 metres. *(Met. Office)*

LEFT A fairly-well-equipped climatological station. The anemometer and the wind vane would be affected by both surface turbulence and by eddies caused by the brick pillars. *(Dave Gavine)*

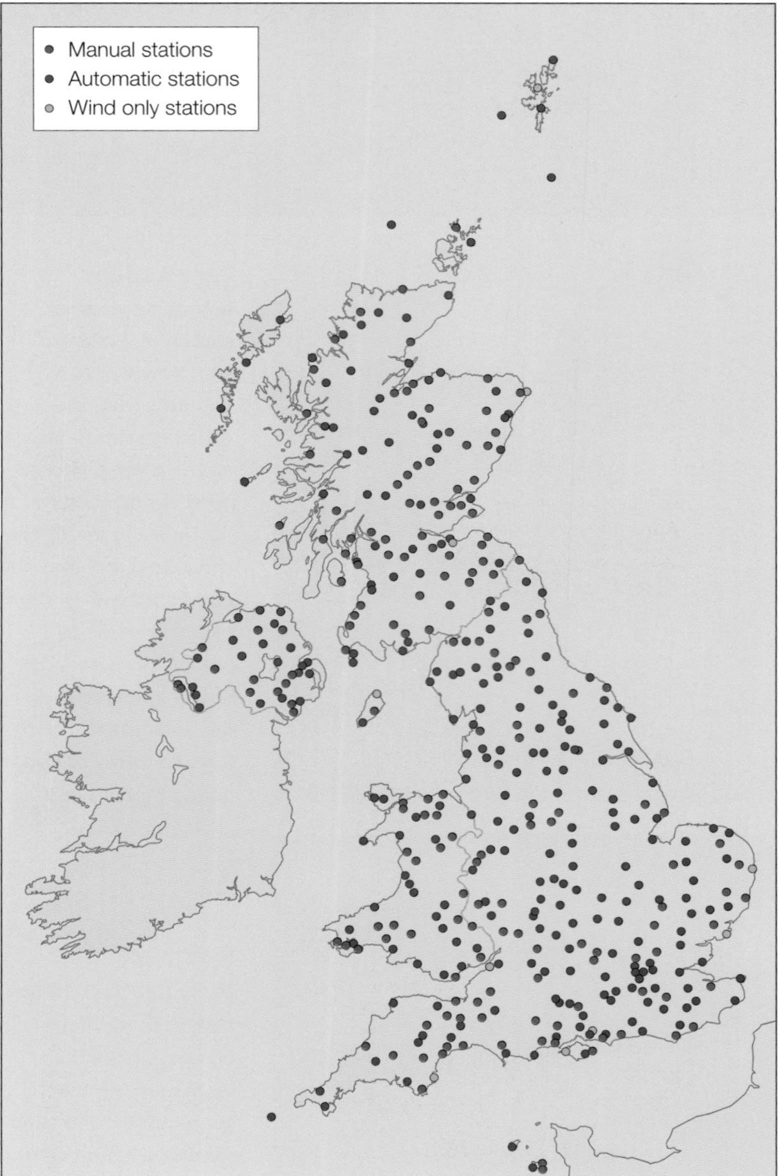

to be carried at such a height, appropriate carefully calculated offsets are applied to all measurements.) Sunshine recorders need to be situated so that they have a clear view from horizon to horizon (*ie* from sunrise to sunset) at all times of the year for their records to have any value. Rain gauges are very difficult to site correctly. Those that are located on the ground may require special protection in highly exposed, windy situations, and generally need precautions to ensure that splashes from the ground cannot enter the collecting funnel.

Many countries are covered by a dense network of stations. Some of these are specifically operated to provide observations for forecasting purposes, whereas the data from many others are used for climatological studies. A few official stations solely provide information on wind strength and direction. In addition, many organisations other than official meteorological offices operate automatic stations for specific purposes. The roadside units operated by highway authorities to monitor road conditions will be familiar to everyone, but other similar stations are used by agricultural and horticultural institutes, water and power authorities, wind farms and other organisations. In certain cases the information from these additional stations may be used to supplement climatological data gathered by the official meteorological organisations. Various forms of automatic weather station are now within the reach of amateurs, and numerous systems have been installed worldwide. Many of these contribute observations to both official and unofficial networks, thus extending coverage of prevailing conditions.

LEFT A map of the official meteorological stations run by the United Kingdom Meteorological office. Note the three types shown: manned, automatic, and wind-only stations. An interactive version of this map may be seen at: http://www.metoffice.gov.uk/public/weather/climate-network/.

RIGHT A professional barograph with an expanded scale to show minor fluctuations in pressure. *(Canadian Met. Service)*

Critical observations for forecasting

The most important factors that are needed to determine forthcoming weather (*ie* for forecasting) are:

- Air pressure (increase or decrease and rate of change).
- Air temperature.
- Air humidity.
- Air movement (direction and strength).

Although many other measurements are made by meteorological and climatological stations, consideration of the instruments used and of the means in which they are employed would be far too extensive to discuss here. Nevertheless, brief descriptions of the way in which these all-important parameters are measured seem in order.

Pressure

The earliest form of barometer was devised by Evangelista Torricelli, a pupil of Galileo Galilei's, in the 17th century. The original design was one of a tube, closed at one end, filled with mercury, and inverted into a basin of mercury. The column of mercury descended in the tube, leaving a vacuum at the closed end, until the weight of the mercury column was balanced by the exterior atmospheric pressure. The oldest existing records of atmospheric pressure are those obtained by Vincenzo Viviani (another pupil of Galileo's, who edited his collected works) and Alfonso Borelli over the period November 1657 to May 1658.

Although many weird and wonderful forms have been devised over the years, the basic design has remained in use until modern times, because although mercury barometers are delicate, they are exceptionally accurate and stable. They and mercury thermometers are being phased out, however, because of concerns over the toxic nature of mercury.

In modern times atmospheric pressure is measured in units (usually given in hectopascals) named after Blaise Pascal, who first definitively established that pressure declined with height.

The most common form of sensing element in barometers, especially domestic ones, consists of an aneroid capsule. This is a partially evacuated metallic capsule, the walls of which are prevented from collapsing by an internal spring. Variations in atmospheric pressure cause the walls to flex, changing the distance between them, and this motion is amplified by

RIGHT A modern reproduction of the barometer devised by Robert Fitzroy, incorporating a thermometer (bottom right) and a so-called 'storm glass' (bottom left) containing liquid that was thought to change appearance with changes in the weather. *(Author)*

BELOW An aneroid barometer. *(LACO Inc)*

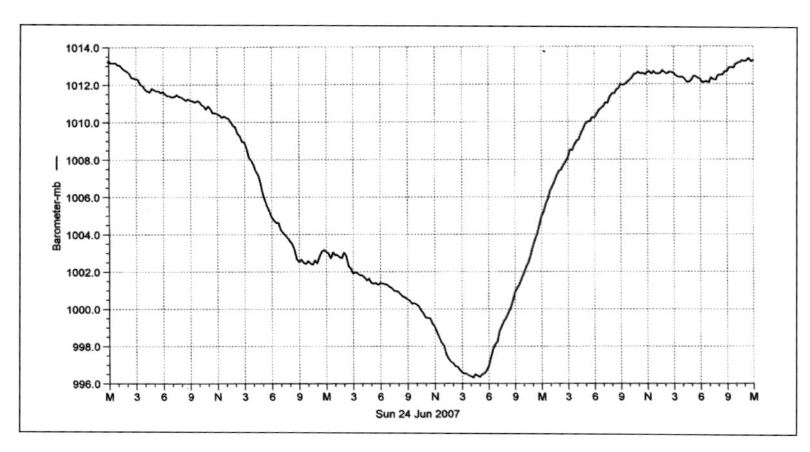

ABOVE The barograph trace of a depression. Pressure fell with the approach of the warm front, at which it steadied slightly, but then continued to fall, until the cold front arrived, heralding a swift rise in pressure. *(Author)*

a suitable system of levers to operate a suitable indicator (as in the common form of household barometer) or to move a pen across a paper chart to provide a permanent recording of pressure changes (as in a barograph). Several capsules may be connected together to amplify the movement. In precision barographs, the exact amount of motion is sensed through the use of an electrical contact.

Electronic pressure sensors detect the movement in an aneroid capsule by electronic means, producing an electrical output that is ideally suited for use in automatic weather stations (AWS) and for subsequent computer processing. Although very robust, such systems tend to be subject to calibration drift and should be recalibrated periodically relative to a reference instrument or, when conditions are appropriate, by the overall local (synoptic) pressure field as determined by official meteorological stations. (Such conditions are occasionally specifically mentioned in radio and television weather forecasts.)

Cross references

*Dew point p.****
*Humidity p.****
*Temperatures and temperature scales p.****

RIGHT The exterior of a Stevenson screen, in this case at a Canadian observing station at Cambridge Bay, Nunavut. Note the fan which draws internal air across the thermometers and thermistors. *(Cambridge Bay Weather)*

Temperature

As with pressure, so the earliest instruments for monitoring temperature were devised in the early 17th century. In 1607 Galileo Galilei produced a device, known as a thermoscope, that displayed changes in temperature by the expansion and contraction of a liquid within a transparent tube. Adding a scale to such a device turned it into a true thermometer. Although such a scale was employed by Robert Fludd in 1638 and an even earlier scale was used by Francesco Sagredo (a close friend of Galileo's) around 1612, the first truly reproducible thermometers were manufactured by the Danish astronomer Ole Rømer – better known for being the first person to determine the speed of light – in 1701.

Once reproducible thermometers were available they became widely used for meteorological studies, although, because of differences in exposure, intercomparison of temperatures and the use of accurate readings for forecasting purposes only became possible with the widespread adoption of suitable temperature screens, typically the Stevenson screen or its equivalent.

Before the introduction of electronic temperature sensors, it was usual for Stevenson screens to contain four thermometers:

- A dry-bulb thermometer, giving the ambient air temperature.
- A wet-bulb thermometer, the bulb of which is kept damp by a wick fed from a container of distilled water. The bulb is cooled by the evaporation of the water, so the thermometer will normally give a lower reading than the dry-bulb thermometer. The difference between the readings may be used (together with knowledge of the ambient pressure) in the calculation of the relative humidity.
- A maximum thermometer, provided with an index that is moved along the tube by the expansion of the liquid within it (usually

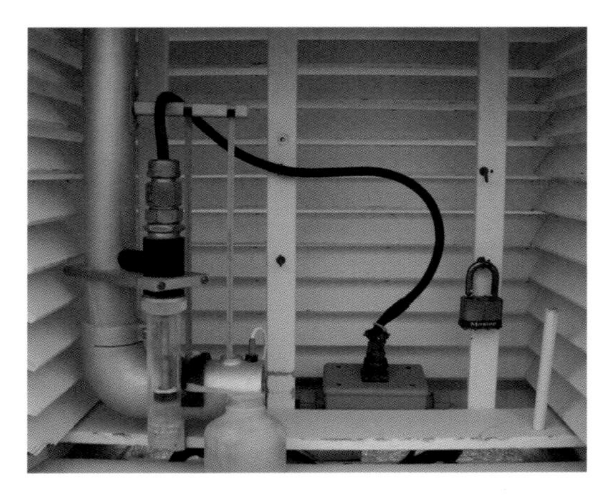

ABOVE The interior of a modern Stevenson screen. The thermometers are aspirated (air is drawn out of the screen through the metal tube). The wet-bulb thermometer (orange) is kept moist by water fed through the wick. The dry-bulb thermometer is yellow. The black cable leads to a thermistor. *(Cambridge Bay Weather)*

LEFT Inside the Stevenson screen at Moree AWS. Manually read dry- and wet-bulb thermometers are placed at the front left. Electronically read thermistors are located at the back of the box. The thermometers placed nearly horizontally measure daily minimum and maximum temperatures. *(Cambridge Bay Weather)*

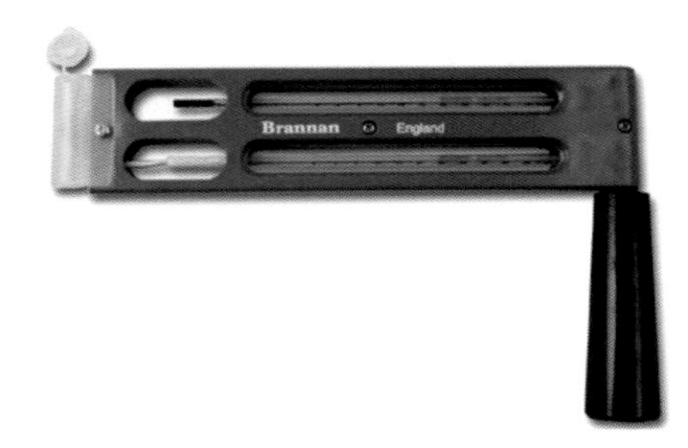

LEFT A whirling psychrometer (also known as a whirling hydrometer), with wet- and dry-bulb thermometers. The distilled-water reservoir for the wet-bulb thermometer (bottom) is visible on the left. *(Brannan)*

mercury), but which remains at the highest point until reset by the observer.

- A minimum thermometer, operating on a similar principle, provided with an index that is drawn down by the contraction of the fluid (usually alcohol), but which remains at the lowest point until reset.

Stevenson screens were often constructed to a larger size to accommodate additional instruments, such as barographs and hygrographs (recording pressure and humidity respectively). Many modern screens incorporate dry or wet thermistors – whose resistance varies with temperature – and thus provide an electrical output that may be recorded by suitable electronics.

Humidity

As just mentioned, the humidity of the air may be determined from the difference between the readings given by dry-bulb and wet-bulb thermometers, given the ambient atmospheric pressure, by the use of what are termed psychrometric tables. For manual measurements, a device known as a whirling psychrometer, with a form reminiscent of a football rattle, may be used. It consists of a frame holding a dry-bulb and a wet-bulb thermometer, that is whirled round. From the two readings the humidity may be derived from the accompanying tables.

Air movement (direction and strength)

As we have seen, winds are generated by pressure differences. Today meteorologists measure wind speeds in metres per second (m/s), although they are also often quoted in knots (for sailors), and kilometres per hour or miles per hour (for the general public). In addition, the 13-point scale proposed by Rear Admiral Beaufort (originally used by him in 1806, but not officially adopted by the British Admiralty until 1838), and initially designed for use at sea, is frequently used to describe the force of the wind. Rather than being defined in terms of speed, the Beaufort Scale is based on the effect of the wind on specific objects. Originally it was defined by its effect on the sails that a typical frigate could carry (frigates being a relatively uniform class of man-of-war at that period, on which all naval officers would have served). Subsequently, especially after the general introduction of steamships, the scale was modified to take account of the state of the sea, and it remains in this form to this day. Later an alternative scale was introduced to describe the effects on land.

Cross references

Enhanced Fujita scale p.121
Saffir-Simpson scale p.126
Tornadoes p.121
Tropical cyclones p.124

As we see elsewhere, a similar evolution occurred with the Fujita Scale that is used to describe the strength of tornadoes. Initially it was based on an assessment of the damage caused, but this proved to be unreliable, given the differences in the strength of construction of various buildings, and in the other criteria that were used. A revised scale (the Enhanced Fujita Scale) is now based on wind speed alone.

In 1946 the Beaufort Scale was extended to Forces 13 to 17, but this classification did not remain in general use, although it is occasionally applied to typhoons in the western Pacific Ocean by some of the meteorological services in the region. Although tropical cyclones (also known as hurricanes, cyclones or typhoons, depending on the ocean in which they occur) are defined as having sustained wind speeds in excess of 33m/s (64 knots, 118kph, 74mph), their severity is nowadays classified on the Saffir-Simpson scale, particularly hurricanes that occur in the North Atlantic.

The direction of the wind may be measured by a simple wind vane, of course, and very early examples are known. The Tower of the Winds in Athens, for example, erected around 50 BC

BEAUFORT SCALE FOR USE AT SEA

Force	Description	Events at sea	m/s	knots	kph	mph
0	Calm	Like a mirror	0.0–0.2	< 1	< 2	< 1
1	Light air	Ripples; no crests or foam	0.3–1.5	1–3	2–6	1–3
2	Light breeze	Small wavelets with smooth crests	1.6–3.3	4–6	7–11	4–7
3	Gentle breeze	Large wavelets; some crests break; a few white horses	3.4–5.4	7–10	12–19	8–12
4	Moderate breeze	Small waves; frequent white horses	5.5–7.9	11–16	20–30	13–17
5	Fresh breeze	Moderate, fairly long waves; many white horses; some spray	8.0–10.7	17–21	31–39	18–24
6	Strong breeze	Some large waves; extensive white foaming crests; some spray	10.8–13.8	22–27	40–50	25–30
7	Near gale	Sea heaping up; streaks of foam blowing in the wind	13.9–17.1	28–33	51–61	31–38
8	Gale	Fairly long and high waves; crests breaking into spindrift; foam in long prominent streaks	17.2–20.7	34–40	62–74	39–46
9	Strong gale	High waves; dense foam in wind; wave-crests topple and roll over; spray interferes with visibility	20.8–24.4	41–47	75–87	47–54
10	Storm	Very high waves with overhanging crests; dense blowing foam, sea appears white; heavy tumbling sea; poor visibility	24.5–28.4	48–55	88–102	55–63
11	Violent storm	Exceptionally high waves may hide small vessels; sea covered in long, white patches of foam; waves blown into froth; visibility severely affected	28.5–32.6	56–63	103–117	64–73
12	Hurricane	Air filled with foam and spray; extremely bad visibility	≥ 32.7	≥ 64	≥ 118	≥ 74

BEAUFORT SCALE FOR USE ON LAND

Force	Description	Events on land	m/s	knots	kph	mph
0	Calm	Smoke rises vertically	0.0–0.2	below 1	below 2	< 1
1	Light air	Direction of wind shown by smoke, but not by wind vanes	0.3–1.5	1–3	2–6	1–3
2	Light breeze	Wind felt on face; leaves rustle; wind-vane turns to wind	1.6–3.3	4–6	7–11	4–7
3	Gentle breeze	Leaves and small twigs in motion; wind extends small flags	3.4–5.4	7–10	12–19	8–12
4	Moderate breeze	Wind raises dust and loose paper; small branches move	5.5–7.9	11–16	20–30	13–17
5	Fresh breeze	Small leafy trees start to sway; wavelets with crests on inland waters	8.0–10.7	17–21	31–39	18–24
6	Strong breeze	Large branches in motion; whistling in telephone wires; difficult to use umbrellas	10.8–13.8	22–27	40–50	25–30
7	Near gale	Whole trees in motion; difficult to walk against wind	13.9–17.1	28–33	51–61	31–38
8	Gale	Twigs break from trees; difficult to walk	17.2–20.7	34–40	62–74	39–46
9	Strong gale	Slight structural damage to buildings; chimney pots, tiles and aerials removed	20.8–24.4	41–47	75–87	47–54
10	Storm	Trees uprooted; considerable damage to buildings	24.5–28.4	48–55	88–102	55–63
11	Violent storm	Widespread damage to all types of building	28.5–32.6	56–63	103–117	64–73
12	Hurricane	Widespread destruction; only specially constructed buildings survive	≥ 32.7	≥ 64	≥ 118	≥ 74

(or possibly even earlier), used a wind vane in the form of a triton to indicate one of eight representations of named winds.

Measurement of the strength of the wind is more difficult. Although Leonardo da Vinci is often said to be the originator of what is known as the swinging-plate anemometer (which is shown in his drawings), in fact the earliest known form was devised by Leone Battista Alberti in 1450.

Nowadays the cup type of anemometer is widely used, together with the windmill type, usually in the form of an aerovane, which combines a propeller with a vane to keep the propeller facing into the wind. These two forms are those generally found in the entry-level and slightly more sophisticated automatic weather stations that are available for amateur use. Another type of anemometer is the hot-wire form, where a wire is heated above ambient temperature and is cooled by the air blowing across it. The drop in temperature thus caused alters the resistance of the wire, providing a suitable electrical signal.

There are numerous other forms. Sonic anemometers use the propagation of ultrasound from a transducer to a receiver to determine the motion of the air between the two, and may be arranged to measure the airflow in three dimensions. Acoustic resonance anemometers determine the wind speed from the performance of a tuned resonant cavity. Laser anemometers rely upon backscattered light from natural or artificial particles within the airstream. One sensitive form is the pressure-tube anemometer, which determines the velocity from the difference in pressure between an open tube, turned to face the wind, and that within a closed tube with apertures to the open air. (A similar principle is used in the pitot tubes employed by aircraft to measure their airspeed.)

Plotting observations

The data from the various instruments (and many other forms of information that have not been mentioned specifically) are plotted for each station using a standardised method and symbols. Such station plots show the overall situation at a specific observational time – they are synoptic plots. A typical plot for the British Isles and a portion of Europe is shown in the diagram.

LEFT The remnants of the Tower of the Winds, visible in Athens today. Two of the carved representations of the eight named winds are clearly visible in the frieze. *(Wikipedia Commons)*

LEFT A typical ultrasonic anemometer, able to determine air motion in three dimensions. The three transducers and receivers are clearly visible. *(Author)*

BELOW A UK Meteorological Office plot of station observations for 15:00 Z on 7 July 2007. The current plot is available at: http://www.metoffice.gov.uk/education/teachers/latest-weather-data-uk *(Met Office)*

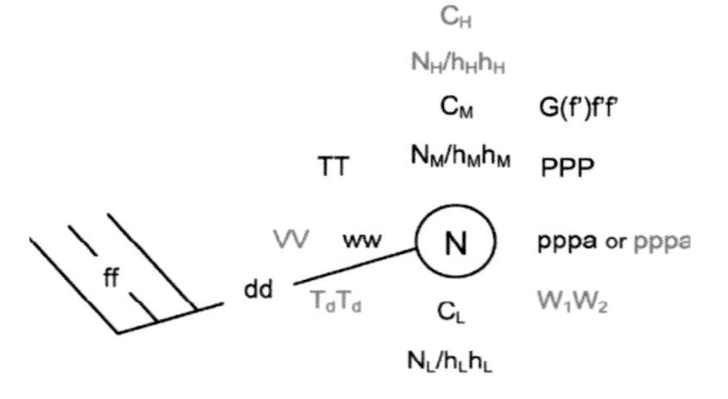

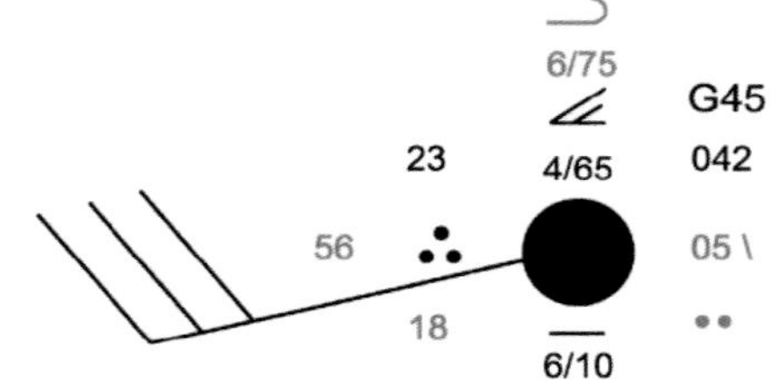

RIGHT The decode (left) for a land station plot, with an actual plot shown (right). *(Met Office)*

Station plots

Such station plots (drawn by computer) are now widely available over the Internet and give a useful view of the weather over a large area at any particular time, the wind, pressure and present weather information being perhaps of the greatest interest. Because of various factors, however, station plots – especially those covering more than one country – have a tendency to be rather variable in content, with the presentation depending on which national meteorological organisation is responsible for producing the charts. The UK Meteorological Office uses different symbols to distinguish between the various types of station (manned, automatic, etc), but some other organisations do not. They may use different basic station symbols, or employ different colours for the plotted data. Perhaps the greatest limitation on the usefulness of such plots is the fact that the stations that are shown may vary greatly from one plot to the next.

In addition, individual measurements may be missing from some of the individual station plots. This may arise because suitable observations were not made, or because there were errors in the transmission of the data to the meteorological office centre. It should be noted, however, that the actual symbols and the codes that are used (and their positioning relative to the station 'circle') are all governed by international agreement, so they remain constant. The example station plots shown here follow the practice of the United Kingdom Meteorological Office.

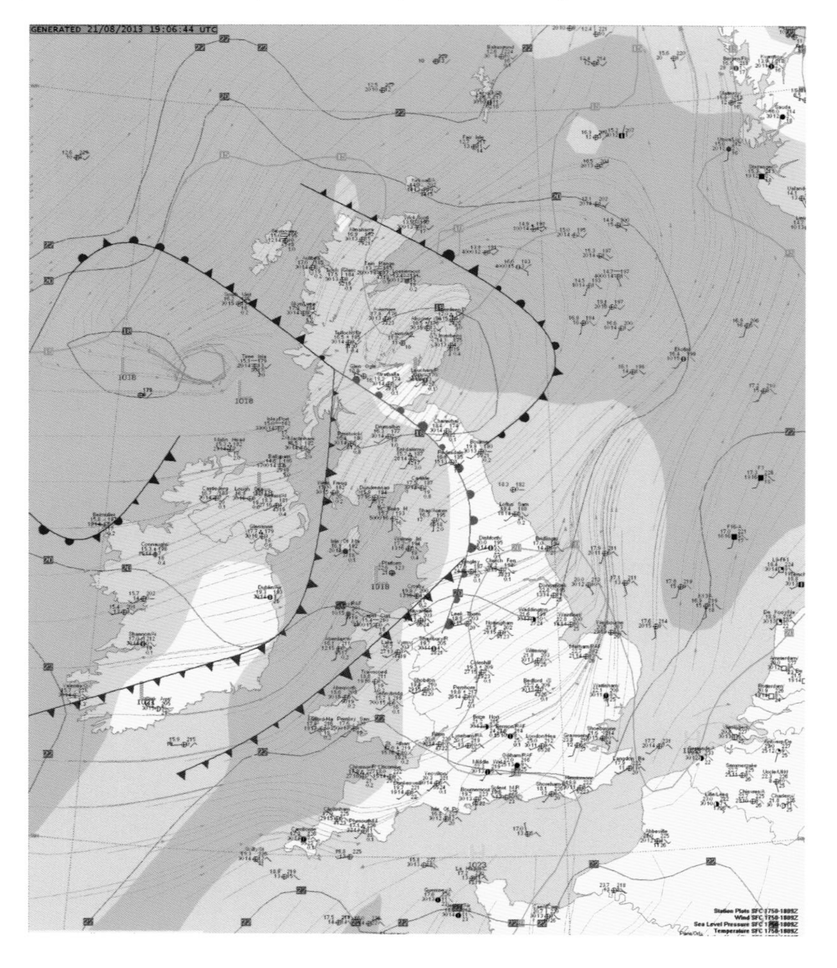

BELOW A sample chart from the Australian online archive of UK charts. The grey lines are streamlines showing airflow (note the arrowheads). The archive is at http://www.australianweathernews.com/sitepages/charts/611_United_Kingdom.shtml. *(Australian Weather News)*

Analysis charts

In addition to producing the station plots, the flood of data disseminated by the Global Telecommunications System is processed by the powerful computers of numerical weather prediction (NWP) systems to produce a model of the state of the atmosphere at that specific time. The information may be presented in visual form as synoptic charts (analysis charts) that in most cases show surface pressure (in the form of isobars), together with frontal systems. Some meteorological services, such as the Australian Weather Service, prepare charts that do not include isobars, but instead combine station plots with a depiction of frontal systems, together with other data.

Forecast charts

The use of powerful supercomputers in recent years has brought a big advance in the production of forecasts. The time span for accurate forecasts has improved from some hours or a day to several days, and

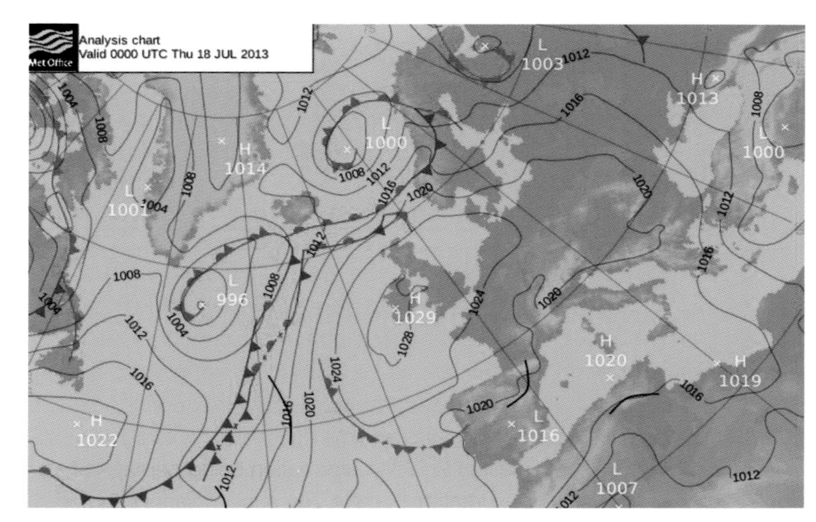

RIGHT A typical analysis chart as prepared by the UK Meteorological Office for 00:00 UTC on 18 July 2013. Current charts may be viewed at http://www.metoffice.gov.uk/public/weather/ surface-pressure/#?\tab=surfacePressure. *(Met Office)*

meteorological services are able to give these with a considerable degree of confidence. The United Kingdom Meteorological Office, for example, routinely gives forecasts for four days ahead. With a gradual refinement of computational methods (and still more powerful computers) the overall accuracy and timescale covered is gradually being extended.

Although the methods vary slightly between individual meteorological services, the basic procedure used by numerical weather prediction consists of employing a set of equations to calculate the changes that will occur in a number of fundamental parameters (such as pressure, temperature, humidity, wind speed and wind direction) over a given time interval within a model of the atmosphere. Generally the time interval is 15 minutes, and the computations are taken forward in such steps for the particular period required for a specific forecast. Typical forecast periods range from a few hours to a few days.

The synoptic reports from observing stations are used to derive appropriate numerical values for the various elements at a set of grid points that are distributed around the globe and at multiple layers in the atmosphere. Ideally, of course, the observing stations would be located in a precisely spaced grid around the world, but as this is obviously not the case the grid-point values have to be suitably interpolated from the known data. This obviously introduces a certain amount of uncertainty in the values that are used as basic input to the various equations. (It was this uncertainty – and other factors – that Edward Lorentz described and which became the basis of chaos theory.)

The methods used by the European Centre for Medium-Range Weather Forecasting (ECMWF) in Reading in the UK may be given as an example of NWP. The predictions by this centre are regarded as some of the most accurate (if not the most accurate) of those that are currently being made. (This international centre is supported by 20 European states, with cooperative agreements with another 14 European states and 7 international organisations.) One of ECMWF's global models has a grid interval of approximately 16km and 91 levels. (Another, newer model will use 137 levels.) The model thus forecasts temperature, humidity and wind for 194,804,064 grid points throughout the atmosphere. The forecast proceeds in 15-minute steps, with the output from one step being used as the input for the next step. Daily, three-day and ten-day forecasts are produced by this method. Forecasts for a longer period ahead use a coarser (80km) grid, as do ensemble forecasts to ten days – which use a different procedure to assess their accuracy. Global wave forecasts use a 55km grid, and wave forecasts for European waters use a closer 27km grid.

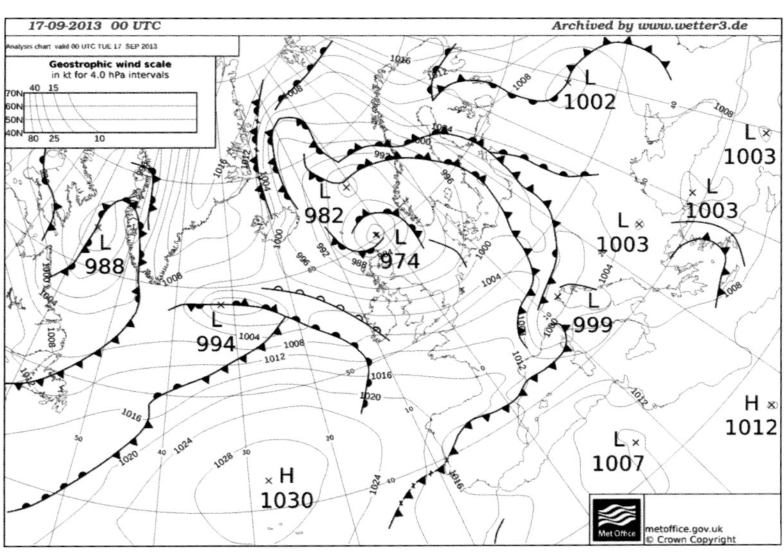

ABOVE A sample UK Meteorological Office analysis chart as available from the German archival site: http://www.wetter3.de/Archiv/archiv_ukmet.html, where records are available since 11 October 2006. *(DWD)*

Chapter 14

Observing techniques

In observing clouds and other phenomena, there are a few simple techniques that are often useful. The first – and it may seem obvious – is to minimise glare from the Sun (or occasionally the Moon). This may be done by hiding the Sun or Moon behind some suitable object – which is essential if one is trying to take a photograph – or by simply blocking the light with your hand. Increasing the contrast between clouds and sky will also help. Ordinary sunglasses may be used, and the mirror-type (mirror shades) are particularly helpful. Viewing the reflections of clouds in a pool of water, or in the dark glass often used for glazing in modern buildings, may also be of considerable assistance. The most effective method, however, is to use polarising material. Try to obtain two small pieces of polariser (usually plastic). One (or a camera polarising filter) may be used on its own and turned to increase the contrast between clouds and sky. Two may be used in combination to decrease the amount of light that they pass, from almost complete transparency to fully blocking the light. (Light reflected from water or other surfaces is often partially polarised, which is one reason that method of viewing may be effective.) Proper polarisers are almost essential when photographing clouds.

One technique, which may seem too simple to be of much use, is to use binoculars to examine clouds. They will, in fact, help you to see how clouds develop, by revealing the motion in the tops of rising cells, for example, and also show subtle details that may otherwise be missed. Using them to examine lenticular clouds will often reveal how the clouds are forming on the upwind edge and dispersing on the downwind side. Similarly, using binoculars may make it easier to decide when a Cumulonimbus cloud is turning into the calvus variety and then on into the capillatus form. Once the latter arrives, you know that the cloud will soon start to produce precipitation. Binoculars will also help to detect what is known as 'jumping Cirrus', a form of Cirrus that surges upwards when part of an overshooting top of a Cumulonimbus cloud suddenly collapses back down into the cloud. Because overshooting tops themselves are only readily visible when you are a considerable distance from the cloud itself, the additional magnification is very useful.

One point to remember, however, is that it is extremely dangerous to use binoculars (or any optical instrument) near the Sun – and certainly never use them to observe the Sun. Even when the Sun is close to the horizon, and appears reddened and greatly attenuated, it is still producing harmful, invisible infrared radiation that may cause damage when concentrated on to the retina by any optical instrument. The Sun may be observed only with proper equipment or by using specifically designed optical filters.

Measuring angles

It is often useful to be able to estimate angles on the sky. The angle subtended by the cloud elements, for example, determines the difference between a layer of cloud being classed as Cirrocumulus, Altocumulus, or Stratocumulus. One method is to hold a rule, graduated in centimetres, at arm's length. One centimetre is then approximately 1° in width. (A 12in rule is slightly more than 300mm, or 30°: a very useful measurement for determining cloud type.) If no rule is available, the hand, again held at arm's length, forms a very convenient method of estimating angles, and one that is surprisingly accurate.

1°	Width of a finger-tip.
2°	Width of the thumb.
7°	Width over four knuckles.
10°	Width of clenched fist.
22°	Span of spread fingers (thumb to little finger).

The angular sizes of some meteorological phenomena are:

0.5°	Diameter of Sun or Moon.
1° or less	Width of Cirrocumulus elements at 30° altitude.
1–5°	Width of Altocumulus elements at 30° altitude.
10° or more	Width of Stratocumulus elements at 30° altitude.
22°	Radius of primary (inner) halo.
30°	Altitude at which cloud types are identified.
42°	Radius of primary rainbow.
46°	Radius of secondary (outer) halo.
52°	Radius of secondary (outer) rainbow.

The radii of rainbows vary to a certain extent, depending on the particular colour considered. The radii of ice-crystal haloes do vary very slightly, depending on colour, but the differences are rarely perceptible.

If observing at night (noctilucent clouds, aurorae, or lunar rainbows or haloes) and the Plough is visible in the northern sky, the separations between its various stars may also be used as a guide to angles.

Amateur weather stations

In recent years many forms of automatic weather station have been produced that are suitable for amateur use, ranging from entry-level stations to advanced types which, if suitably installed and calibrated, are able to produce results that are acceptable as data input to the official, national meteorological services.

All these systems (even the most inexpensive entry-level ones) generally record temperatures, humidity, pressure, rainfall, wind speed and direction. These are all essential factors in the production of weather forecasts. Naturally, the more expensive systems may be expected to be more accurate, offer more consistent measurements, and perhaps have a longer lifetime than the cheaper entry-level systems.

In any 'domestic' situation, the greatest problem is likely to be the siting of the system. Except in a very few locations, mounting the anemometer and wind vane at the ideal

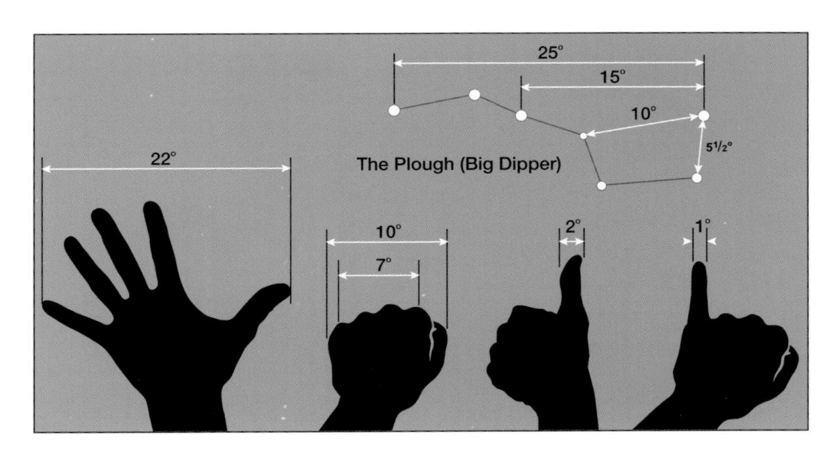

ABOVE Holding the hand at arm's length is a useful way of obtaining approximate angles on the sky. If it is a clear night, the distances between stars of the Plough may also be used. *(Ian Moores)*

LEFT An inexpensive 'entry-level' automatic weather station, with a wireless link to the touch-screen display, which may be connected to a computer. *(Maplin)*

LEFT An entry-level AWS system installed. Note the size and shape of the rain gauge (the small rectangular plastic box). *(Author)*

RIGHT A 'budget' AWS system (a Davis Instruments Vantage Vue). Although not visible here, the rain gauge has a circular-aperture, plastic hopper on the top of the unit. *(Davis Instruments)*

RIGHT A 'semi-professional' AWS system (Davis Instruments Vantage Pro2), available in cabled or wireless forms. The larger, circular-aperture rain gauge is of a more satisfactory design. *(Davis Instruments)*

(and recommended) height of 10m is likely to be quite impractical. At a lower height, the measurements will probably be severely affected by turbulence caused by buildings, trees, garden walls and other obstacles. Any measurements are likely to show wide variations, and give just a general indication of actual conditions.

All automatic weather stations of the amateur and semi-professional type incorporate some form of tipping-bucket rain gauge, and it seems to be the general experience that this is the least effective piece of instrumentation, even in some of the 'semi-professional' designs. More consistent results are given by the traditional ('Snowdon') form of rain gauge with a circular collecting aperture having a very accurately machined, knife-edge rim, the diameter (and thus area) of which is precisely determined. Rain amounts are read manually using a graduated measuring cylinder.

Ideally, of course, any such systems should be calibrated against instruments that are known to be accurate, and such a calibration should be carried out at intervals to compensate for any drift in the sensors or electronics. The pressure reading may be adjusted on those occasions when radio or television weather forecasters suggest that conditions are ideal for setting barometers. Temperature calibration has been made more difficult recently by the ban on mercury in barometers and thermometers. The design known as 'Six's thermometer' which showed current temperature and had indexes to show maximum and minimum values recorded since it was last reset, was of reasonable accuracy, but relied upon the use of a mercury column. Attempts to find a single liquid as a substitute appear to be unsuccessful. The 'official' thermometers used to record minimum temperatures used alcohol rather than mercury, and were designed to be accurate at low temperatures. But some form of calibration is necessary if the observer is serious about obtaining accurate readings. Temperature sensors have been found to read 'high' or 'low' over certain portions of their nominal range, and humidity measurements are sometimes suspect. In one case, thick fog meant that the sensor unit was difficult to see from a distance of some 20m. The temperature was obviously below the dew point, yet the relative humidity was displayed as 91%, rather than 100%.

Various software programs are available to display and archive data from automatic weather stations, and some include templates for the display of data on an observer's personal website. Data may also be submitted to various organisations, most notably to the Weather on the Web (WOW) website, hosted by the UK Meteorological Office. Amateur observations are also collated by the Climatological Observers Link (COL), which produces monthly summaries of data.

RIGHT A 'classic' Snowdon rain gauge with a precisely machined, knife-edge rim. *(Author)*

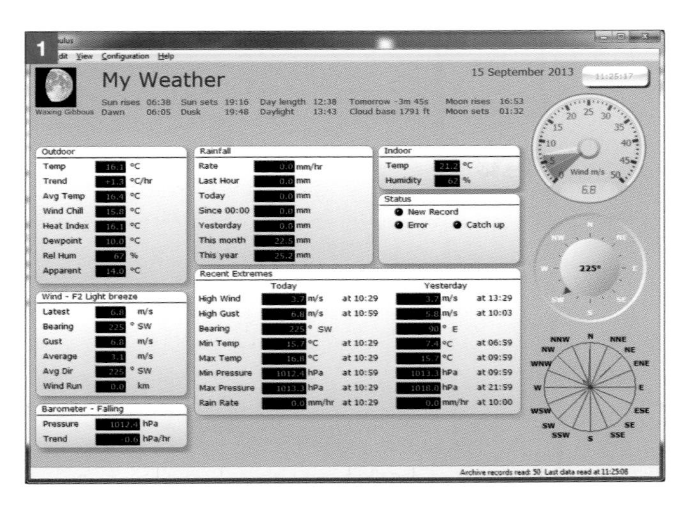

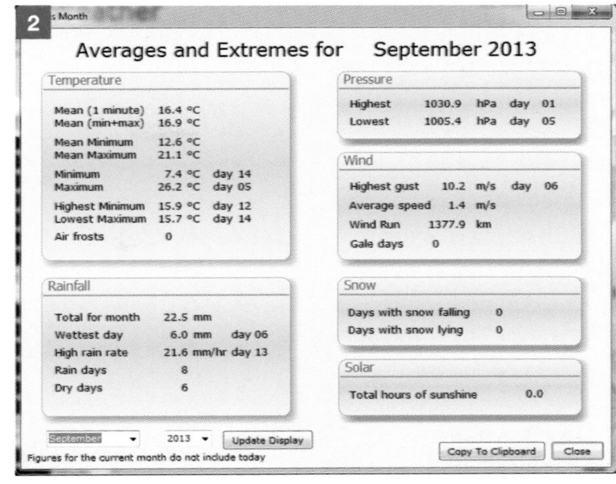

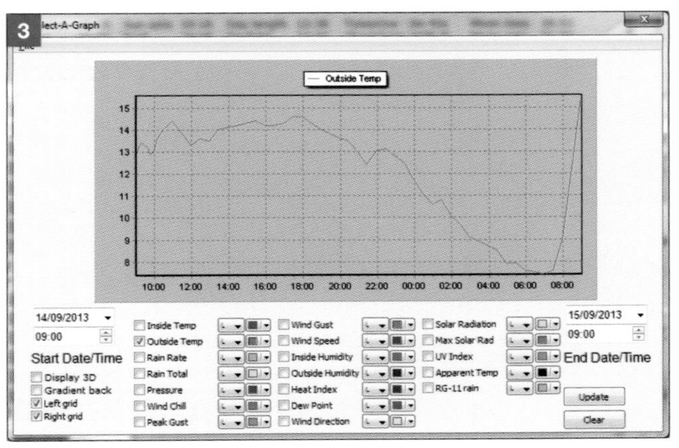

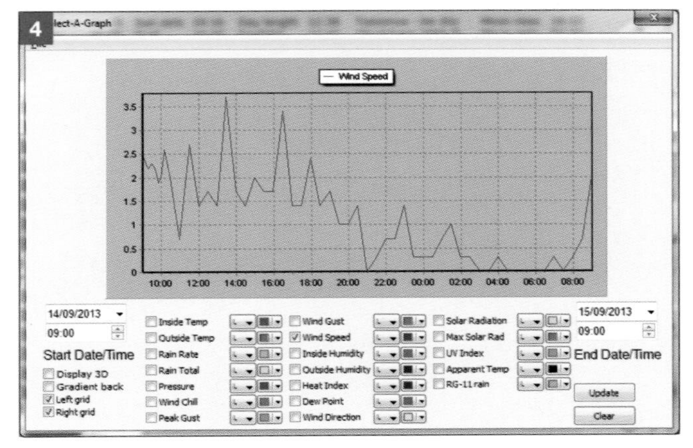

1 **A typical display of data from an AWS system (in this case using the 'Cumulus' software.** *(Cumulus software)*

2 **Averages and extremes for the first half of September 2013.** *(Cumulus software)*

3 **Daily wind speed from 09:00 (UT), 14 September to 09:00, 15 September 2013.** *(Cumulus software)*

4 **Daily outside temperature from 09:00 (UT), 14 September to 09:00, 15 September 2013.** *(Cumulus software)*

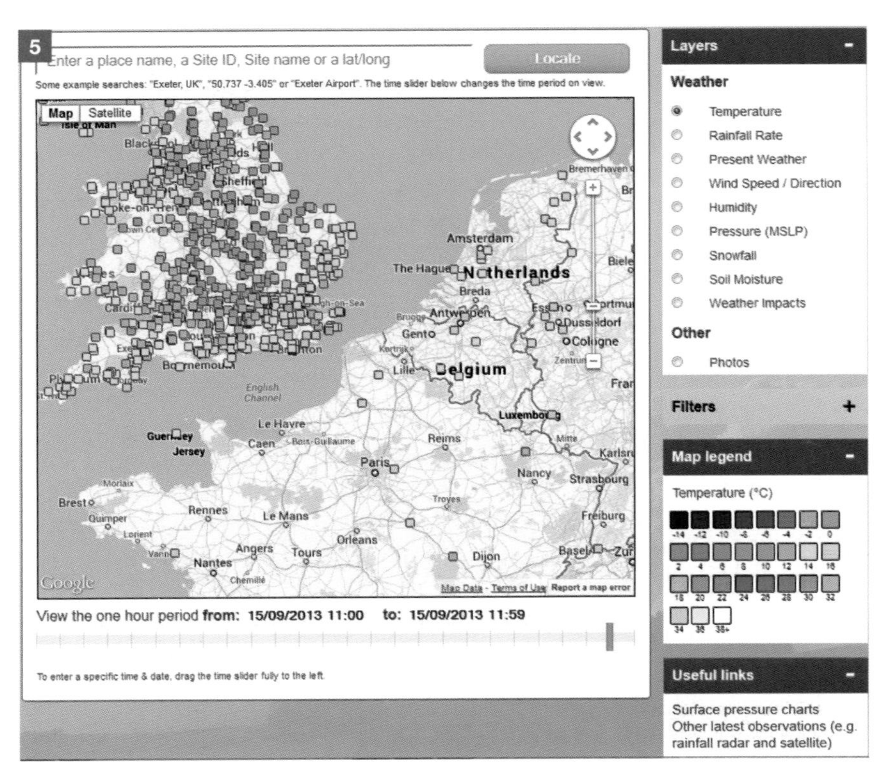

5 **A typical screen display from the Meteorological Office's 'Weather on the Web' (WOW) website at http://wow.metoffice.gov.uk.** *(Met Office)*

PART FOUR

Appendices

OPPOSITE A sheet of Altocumulus stratiformis, behind a depression that had passed away to the east (left), luridly illuminated by the setting Sun, photographed in early March. *(Author)*

Appendix 1

Glossary

Adiabatic – Without the addition or loss of heat. Parcels of air in the atmosphere generally rise and fall without the exchange of heat with their surroundings.
Aerosol – Any minute solid or liquid particle suspended in the atmosphere.
Air mass – A body of air that has acquired specific characteristics, specifically temperature and humidity, by remaining stationary over an area of the globe for some time. It tends to retain its temperature and humidity when it eventually moves away from the source area, and largely determines the weather of any region it crosses.
Anabatic – Moving upwards. The term is typically applied to winds (such as a valley wind) or to the air at frontal systems.
Anticyclone – A high-pressure region that is a source of air that has subsided from higher altitudes, and from which air flows out over the surrounding area. The circulation around an anticyclone is clockwise in the northern hemisphere.
Anticyclonic – Moving or curving in the same direction as air circulating around an anticyclone, *ie* clockwise in the northern hemisphere, anticlockwise in the southern.
Antisolar point – The point on the sky directly opposite the position of the Sun.
Backing – An anticlockwise change in the wind direction, *ie* from west, through south, to east.
Beaufort Scale – A numerical scale for the description of wind speed, ranging from 0: calm; 1: 1–3 knots (0.3–1.5m/s or about 1–3mph) to 12: above 64 knots (above 33m/s or above about 73mph).
Celsius – The correct term for the temperature scale where the freezing and boiling points of water are 0°C and 100°C respectively. (Frequently, and incorrectly, called 'Centigrade'.)
Col – An area of slack atmospheric pressure, located between a pair of low-pressure centres and a pair of high-pressure ones. Slight changes in pressure may cause rapid motion of a col or its disappearance.
Continental climate – A climate that is typical of continental interiors, characterised by extremely cold winters and hot summers. There is also a tendency towards low annual precipitation totals.
Convection – Transfer of heat by the motion of parcels of a fluid such as air or water. In the atmosphere this motion is predominantly vertical. There are two forms of convection: 'natural convection' in which parcels or 'bubbles' of air are free to move vertically driven by buoyancy effects; and 'forced convection' in which air is mixed mechanically by eddies.
Coriolis force – The apparent force, caused by the rotation of the Earth, which deflects any moving object (such as a parcel of air) away from a straight-line path. In the northern hemisphere it acts towards the right, and in the southern to the left. It increases in proportion to the velocity of the moving object.
Cyclone – A system in which air circulates around a low-pressure core, with two distinct meanings: 1) a 'tropical cyclone', a self-sustaining tropical storm, also known as a hurricane or typhoon; 2) an 'extratropical cyclone' or depression, a low-pressure area, which is one of the principal weather systems in temperate regions.
Cyclonic – Moving or curving in the same direction as air that flows around a cyclone, *ie* anticlockwise in the northern hemisphere, clockwise in the southern.
Depression – The most frequently used term for a low-pressure area. Air flows into a depression and rises in its centre. Known technically as an 'extratropical cyclone'. The wind circulation around a depression is cyclonic (anticlockwise in the northern hemisphere).
Dew point – The temperature at which a particular parcel of air, with a specific humidity, will reach saturation. At the dew point, water

vapour will begin to condense into droplets, giving rise to a cloud, mist or fog, or depositing dew on the ground.

Föhn wind – A hot, dry (and often desiccating) wind that descends on the leeward side of mountains. Having deposited most of its moisture on the windward slopes, it is much warmer and drier than at comparable levels on the opposite side of the range.

Geostrophic – A term applied to a (hypothetical) wind that flows parallel to the isobars. The wind at low-cloud height (600m, 2,000ft) corresponds approximately to a geostrophic wind.

Hurricane – One of several names for a potentially destructive tropical cyclone, used in the North Atlantic and eastern Pacific.

Instability – The condition under which a parcel of air, if displaced upwards or downwards, tends to continue (or even accelerate) its motion. The opposite is stability.

Inversion – An atmospheric layer in which temperature remains constant or increases with height.

Isobar – A line that joins points on a weather chart that have the same barometric pressure.

Jet stream – A narrow band of high-speed winds that lies close to a break in the level of the tropopause, with two main jet streams (the Polar Front and Subtropical jet streams) in each hemisphere. Other jet streams exist in the tropics and at higher altitudes.

Katabatic – Moving downwards. Used primarily in connection with katabatic winds (fall winds) that sweep down from high ground, and are normally initiated by low temperatures over the higher ground. The term is also applied to the motion of the air at certain frontal systems.

Kelvin – A unit of heat (K), used to express a temperature scale that begins at absolute zero (-273.15°C). Temperatures are expressed in Kelvin units (*eg* 300K), not as 'degrees Kelvin'.

Lapse rate – The rate at which temperature changes with increasing height. By convention, the lapse rate is positive when the temperature decreases, and negative when it increases with height.

Latent heat – The heat that is released when water vapour condenses into droplets or freezes into ice crystals. It is the heat that was originally required for the process of evaporation or melting.

Maritime climate – A climate that is strongly influenced by the region's proximity to the ocean. Generally characterised by significant amounts of precipitation throughout the year, but with generally mild winters and summers that rarely reach extremely high temperatures.

Mock sun – A halo effect consisting of a bright point of light, often slightly coloured and with a white tail. It lies at the same altitude as the Sun and approximately 22° away from it. Known technically as a parhelion.

Mountain wind – A wind that blows down the length of a valley at night, primarily driven by the fact that the air over areas at a higher altitude cools quicker than the air in the more protected valley.

Mesoscale – Atmospheric phenomena that are between approximately 80 and 250km across. They are thus too small to be adequately studied by synoptic charts, but may be investigated by satellite instrumentation and radar systems.

Mesosphere – The atmospheric layer above the stratosphere, in which temperature decreases with height, reaching the atmospheric minimum at the mesopause, at an altitude of either 86 or 100km (depending on season and latitude).

Occluded front – A front in a depression system, where the warm air has been lifted away from the surface, having been undercut by cold air. The front may, however, remain a significant source of cloud and precipitation.

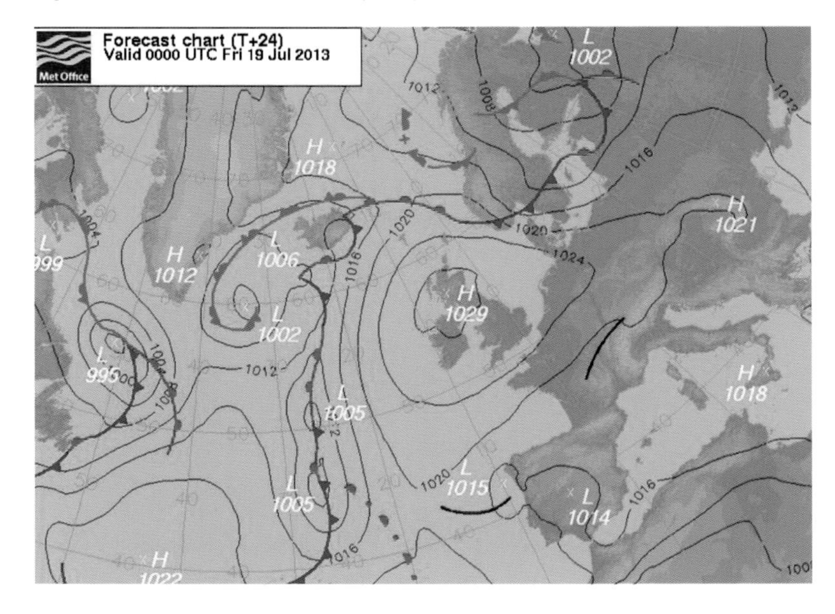

BELOW **A UK Meteorological Office forecast chart, based on the situation shown on the analysis chart shown on page 157, for 24 hours ahead, i.e., for 00:00 UTC on 19 July 2013.** *(Met Office)*

Parhelion – The technical term for a mock sun.
Precipitation – The technical term for water in any liquid or solid form that is deposited from the atmosphere, and which falls to the ground. It excludes cloud droplets, mist, fog, dew, frost and rime, as well as virga.
Pressure tendency – The change in atmospheric pressure during the previous three hours.
Relative humidity – The amount of moisture in the air, normally given as a percentage of the amount that the air would contain if fully saturated at a given temperature.
Ridge – The extension of an area of high pressure, resulting in approximately V-shaped isobars pointing away from the pressure centre.
Stability – The condition under which a parcel of air, if displaced upwards or downwards, tends to return to its original position rather than continuing its motion.
Stratosphere – The second major atmospheric layer from the ground, in which temperature initially remains constant, but then increases with height. It lies between the troposphere and the mesosphere, with lower and upper boundaries of approximately 8–20km (depending on latitude) and 50km respectively.
Supercooling – The conditions under which water may exist in a liquid state, despite being at a temperature below its nominal freezing point, 0°C. This occurs frequently in the atmosphere, often in the absence of suitable freezing nuclei. Supercooled water freezes spontaneously at a temperature of -40°C (-40°F).

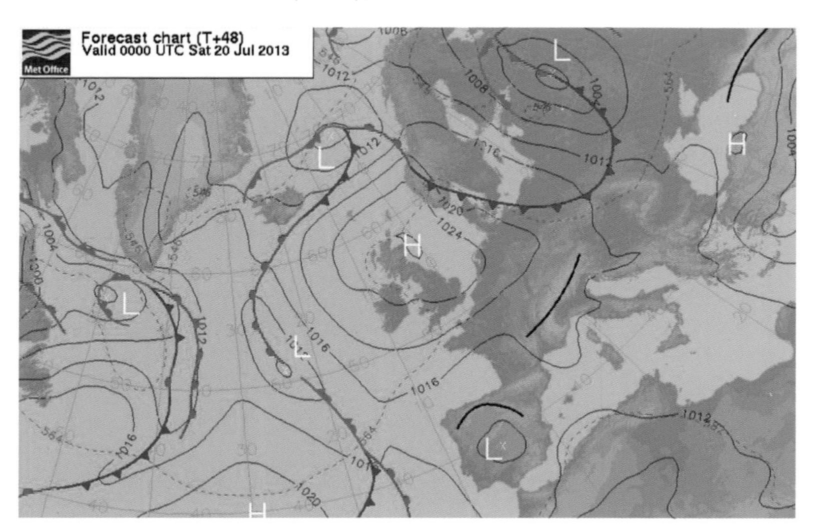

BELOW A UK Meteorological Office forecast chart, based on the situation shown on the analysis chart shown on page 157, for 48 hours ahead, i.e., for 00:00 UTC on 20 July 2013. *(Met Office)*

Synoptic chart – A chart showing the values of a given property (such as temperature, pressure, humidity etc) prevailing at different observing sites at a single specific time.
Synoptic scale – Weather phenomena that are approximately 200–2,000km across, thus lying between mesoscale and planetary scale phenomena in size.
Thermal – A rising bubble of air that has broken away from the heated surface of the ground. Depending on circumstances, a thermal may rise until it reaches the condensation level, at which its water vapour will condense into droplets, giving rise to a cloud.
Tropopause – The inversion that separates the troposphere from the overlying stratosphere. Its altitude varies from approximately 8km at the poles to 18–20km over the equator.
Troposphere – The lowest region of the atmosphere in which most of the weather and clouds occur. Within it, there is an overall decline in temperature with height.
Trough – An elongated extension of an area of low pressure, which results in a set of approximately V-shaped isobars, pointing away from the centre of the low.
Valley wind – A wind that blows up a valley during the day, driven by the greater heating of the air above the upper slopes, which draws air up from lower levels. Its counterpart is the night-time mountain wind.
Veering – A clockwise change in the wind direction, *ie* from east, through south, to west.
Virga – Trails of precipitation (as ice crystals or raindrops) from clouds that do not reach the ground, melting and evaporating in the drier air between the cloud and the surface.
Wind chill – The loss of heat from the skin caused by the effects of wind. Even a moderate wind will create a heat loss that is as great as that occurring at a much lower temperature under calm conditions.
Wind shear – A change in wind direction or strength with a change of position. If, for example, the wind strength increases with increasing height, this is defined as vertical wind shear. If the wind strength changes with motion at a particular level, this is known as horizontal wind shear.
Zenith – The point on the sky directly above the observer's head.

Appendix 2

Useful contacts and further information

Further reading

Identification, general and photographic studies

Chaboud, René. *How Weather Works* (Thames & Hudson, 1996).

Dunlop, Storm. *Collins Gem: Weather* (HarperCollins, 1999).

— *Collins Nature Guide: Weather* (HarperCollins, 2004).

— *Dictionary of Weather* (2nd edition, Oxford University Press, 2008).

— *Guide to Weather Forecasting* (revised printing, Philip's, 2013).

— *How to Identify Weather* (HarperCollins, 2002).

— *Weather* (Cassell Illustrated, 2006/2007).

Eden, Philip. *Weatherwise* (Macmillan, 1995).

File, Dick. *Weather Facts* (Oxford University Press, 1996).

Hamblyn, Richard, and Meteorological Office. *The Cloud Book: How to Understand the Skies* (David & Charles, 2009).

— *Extraordinary Clouds* (David & Charles, 2009).

Ludlum, David. *Collins Wildlife Trust Guide: Weather* (HarperCollins, 2001).

Watts, Alan. *Instant Weather Forecasting* (Adlard Coles Nautical, 2000).

— *Instant Wind Forecasting* (Adlard Coles Nautical, 2001).

Whitaker, Richard (editor). *Weather: The Ultimate Guide to the Elements* (HarperCollins, 1996).

Williams, Jack. *The AMS Weather Book: The Ultimate Guide to America's Weather* (University of Chicago Press, 2009).

More specialised works

Boyer, Carl. *The Rainbow: From Myth to Mathematics* (Princeton University Press, 1987).

Brettle, Mike, and Smith, B. *Weather to Sail* (Crowood Press, 1999).

Burt, Christopher. *Extreme Weather* (W.W. Norton, 2007).

Burt, Stephen. *The Weather Observer's Handbook* (Cambridge University Press, 2012).

Dunlop, Storm. *Photographing Weather* (Photographers' Institute Press, 2007).

Greenler, Robert. *Rainbows, Halos, and Glories* (Cambridge University Press, 1980).

Harding, Maria. *Weather to Travel* (3rd edition, Tomorrow's Guides, 2001).

Henson, Robert. *The Rough Guide to Weather* (2nd edition, Rough Guides, 2007).

Kington, John. *Climate and Weather* (HarperCollins, 2010).

Können, G.P. *Polarized Light in Nature* (Cambridge University Press, 1985).

Meinel, Aden, and Meinel, Marjorie. *Sunsets, Twilight and Evening Skies* (Cambridge University Press, 1983).

Minnaert, M.G.J. *Light and Color in the Outdoors* (Springer, 1993).

Pedgley, David. *Mountain Weather* (3rd edition, Cicerone Press, 2006).

Meteorological Office. *Cloud Types for Observers* (HMSO, 1982).

— *Observer's Handbook* (HMSO, 1982).

Scorer, Richard. *Cloud Investigation by Satellite* (Ellis Horwood, 1986).

Journals

Weather, Royal Meteorological Society, Reading, UK (monthly).

Weatherwise, Heldref Publications, Washington DC, USA (bimonthly).

Internet links

Current weather

AccuWeather: www.accuweather.com.
 UK: www.accuweather.com/ukie/index.asp.
Australian Weather News: www.australianweathernews.com.
 UK station plots: www.australianweathernews.com/sitepages/charts/611_United_Kingdom.shtml.
BBC Weather: www.bbc.co.uk/weather.
CNN Weather: www.cnn.com/WEATHER/index.html.
Intellicast: http://intellicast.com.
ITV Weather: www.itv-weather.co.uk.
Unisys Weather: http://weather.unisys.com.
UK Meteorological Office: www.metoffice.gov.uk.
 Forecasts: www.metoffice.gov.uk/weather/uk/uk_forecast_weather.html.
 Hourly Weather Data: www.metoffice.gov.uk/education/teachers/latest-weather-data-uk.
 Latest station plot: www.metoffice.gov.uk/data/education/chart_latest.gif.
 Surface pressure charts: www.metoffice.gov.uk/public/weather/surface-pressure.
 Explanation of symbols on pressure charts: www.metoffice.gov.uk/guide/weather/sysmbols#pressure-symbols.
 Synoptic & climate stations (interactive map): www.metoffice.gov.uk/public/weather/climate-network/#?tab=climateNetwork.
 Weather on the Web: http://wow.metoffice.gov.uk.
The Weather Channel: www.weather.com/twc/homepage.twc.
Weather Underground: www.wunderground.com.
Wetterzentrale: www.wetterzentrale.de/pics/Rgbsyn.gif.
Wetter3 (German site with global information): www.wetter3.de.
 UK Met. Office chart archive: www.wetter3.de/Archiv/archiv_ukmet.html.

RIGHT Cumulus fractus beneath a large Cumulonimbus cloud, with encroaching Cirrus far above. *(Author)*

FAR LEFT **An overhanging anvil of Cumulonimbus and a distant heavy bank of Cumulus mediocris on a showery day.** *(Author)*

LEFT **Although this sheet of Stratocumulus had a fairly even, featureless base, from above it shows considerable differences in thickness, with convective cloud breaking through in the distance.** *(Author)*

General information

Atmospheric Optics: www.atoptics.co.uk.
Hurricane Zone Net: www.hurricanezone.net.
National Climate Data Centre: www.ncdc.noaa.gov.
  Extremes: www.ncdc.noaa.gov/oa/climate/severeweather/extremes.html.
National Hurricane Center: www.nhc.noaa.gov.
Reading University (Roger Brugge): www.met.reading.ac.uk/~brugge/index.html.
UK Weather Information: www.weather.org.uk.
Unisys Hurricane Data: http://weather.unisys.com/hurricane/atlantic/index.html.
WorldClimate: www.worldclimate.com.

Meteorological offices, agencies and organisations

Environment Canada: www.msc-smc.ec.gc.ca.
European Centre for Medium-Range Weather Forecasting (ECMWF): www.ecmwf.int.
European Meteorological Satellite Organisation: www.eumetsat.int/website/home/index.html.
Intergovernmental Panel on Climate Change: www.ipcc.ch.
National Oceanic and Atmospheric Administration (NOAA): www.noaa.gov.
National Weather Service (NWS): www.nws.noaa.gov.
UK Meteorological Office: www.metoffice.gov.uk.
World Meteorological Organisation: www.wmo.int/pages/index_en.html.

Satellite images

Eumetsat: www.eumetsat.de.
  Image library: www.eumetsat.int/website/home/Images/ImageLibrary/index.html.
Group for Earth Observation (GEO): www.geo-web.org.uk.
University of Dundee: www.sat.dundee.ac.uk.

Societies

American Meteorological Society: www.ametsoc.org/AMS.
Australian Meteorological and Oceanographic Society: www.amos.org.au.
Canadian Meteorological and Oceanographic Society: www.cmos.ca.
Climatological Observers Link (COL): https://colweather.ssl-01.com.
European Meteorological Society: www.wmwtsoc.org.
Irish Meteorological Society: www.irishmetsociety.org.
National Weather Association, USA: www.nwas.org.
New Zealand Meteorological Society: www.metsoc.org.nz.
Royal Meteorological Society: www.rmets.org.
TORRO: Tornado and Storm Research Organisation: http://torro.org.uk.

Index